Lecture Notes in Mathematics 1523

Weimin Xue

Rings with Morita Duality

Springer-Verlag
Berlin Heidelberg New York
London Paris Tokyo
Hong Kong Barcelona
Budapest

Autor

Weimin XUE
Department of Mathematics
Fujian Normal University
Fuzhou, Fujian 350007
People's Republic of China

Mathematics Subject Classification (1991): 16-02, 16D90, 16D50, 16P20, 16P40, 16S20, 16U80

ISBN 3-540-55770-9 Springer-Verlag Berlin Heidelberg New York
ISBN 0-387-55770-9 Springer-Verlag New York Berlin Heidelberg

Printed in Germany

Typesetting: Camera ready by author/editor
46/3140-543210 - Printed on acid-free paper

This book is dedicated to
Kent R. Fuller

TABLE OF CONTENTS

PREFACE

As a generalization of the duality of vector spaces over division rings, Azumaya [59] and Morita [58] established the theory of Morita duality. Such a duality is an additive contravariant category equivalence between two categories of R-left- and S-right-modules, which are both closed under sub- and factor modules and contain all finitely generated modules. Azumaya [59] and Morita [58] have shown that these dualities are precisely those equivalent to the functors Hom(-, E) induced by bimodules ${}_RE_S$ that are injective cogenerators on both sides and satisfy $S = \mathrm{End}({}_RE)$ and $R = \mathrm{End}(E_S)$, and that the natural domain and range of such a duality are the categories of E-reflexive modules, i.e., linearly compact modules by a result of Muller [70]. This book gives an extensive introduction to Morita duality, and includes many important results that have been obtained by various authors. It also presents some updated results on duality.

Chapter 1 contains some basic characterizations of Morita duality. Some of them are taken from Anderson and Fuller [74, Sections 23, 24]. The other characterizations via linear compactness are basically due to Müller [70], whose conjecture that linearly compact commutative rings have dualities has been proved by Ánh [90].

In Chapter 2, we consider Morita duality and ring extensions. Lemonnier [84] proved that a finite triangular extension $S \geq R$ over a ring R with a duality has itself a duality. If R has a self-duality and $S \geq R$ is a finite normalizing extension, Mano [84] showed that, under some additional conditions, S has a self-duality. Kraemer [90] has proved that a finite normalizing extension over a division ring has self-duality. Trivial extensions $R \propto M$ of a ring R by a R-bimodule M are also examined. This is due to Müller [69]. Faith's theorem [79] on PF-rings and trivial extensions is also presented.

In Chapter 3, we consider artinian rings with a duality. It contains well-known theorem of Azumaya [59] and Morita [58]. Fuller's Theorem [69] is also included, which states that if R_e is injective, where $e = e^2 \in R$, then ${}_{eRe}eRf_{fRf}$ defines a duality for some $f = f^2 \in R$. In the conclusion of this chapter we briefly give an introduction to quasi-Frobenius (QF-) rings. Details about QF-rings can be found in Faith [76, Chapter 24].

Azumaya [83] initiated the study of exact rings and conjectured the self-duality for this class of rings. In Chapter 4, we present updated results about Azumaya's conjecture. This is true for serial rings (Dischinger and Müller [84], Waschbüsch [86]). Partial results will be given for locally distributive rings (Fuller and Xue [91], Belzner [90]) and Artinian duo rings (Xue [89], Habeb [89]).

In the final Chapter 5, duality for noetherian rings are studied. Most results are due to Müller [69], Jategaonkar [81], and Menini [86]. As a generalization of perfect rings, Camillo and Xue [91] introduced quasi-perfect rings. The behavior of such rings with duality is considered.

I express my sincere thanks to my PH.D. adviser, Professor Kent R. Fuller (University of Iowa), for his encouragement and invaluable advice. In particular, I thank him for allowing me to use his Unpublished Lecture Notes [F2]. In fact, some of the proofs in Section 11 are taken from there or from his book [89] which is a supplement to Anderson and Fuller [74].

I also wish to express my gratitude to Professors Zhaomu Chen, Luosheng Huang, and Yonghua Xu, for helpful suggestions.

Finally, I thank my wife Danyang Chen and our son Jitian Xue for all things nonmathematical.

Weimin Xue
Fuzhou, Fujian, P.R. China
February 1, 1992

The author is supported by National Science Foundation of China

CHAPTER 1
INTRODUCTION TO MORITA DUALITY

This chapter contains some basic characterizations of Morita duality. In Section 1 we collect some basic results in the theory of rings and modules. Section 2 is a basic introduction to Morita duality and most results in this section are taken from the standard ring theory book, Anderson-Fuller [74, Sections 23, 24]. Sections 3 and 4 are characterizations of this duality via linear compactness. These are basically due to Müller [70], whose conjecture that linearly compact commutative rings have dualities has been proved by Ánh [90]. This result is included in Section 6, and Section 5 contains some basic results of cogenerator rings.

SECTION 1. PRELIMINARIES

In this book, a ring means an associative ring with identity $1 \neq 0$, modules are unitary, and ring extensions share the same identity. If R and S are rings then ${}_RM$ (M_S, ${}_RM_S$) denotes a left R-module (right S-module, a left R-right S-bimodule). Sometimes we just write M to denote a module over a fixed ring if it is not necessary to indicate "left" or "right". If ${}_RM$ is a left R-module and $S = \mathrm{End}({}_RM)$, then ${}_RM_S$ becomes a bimodule in a natural way; i.e., we define $ms = s(m)$ for $m \in M$ and $s \in S$. If N is a submodule (proper submodule, subset, proper subset) of a module (module, set, set) M, we denote $N \leq M$ ($N < M$, $N \subseteq M$, $N \subset M$), or $M \geq N$ ($M > N$, $M \supseteq N$, $M \supset N$).

The results in this section can be found in standard texts on algebra and ring theory, for example, Anderson-Fuller [74], Faith [76], Goodearl [76], Kasch [82], Rotman [79], Sharpe-Vámos [72].

Let R be a ring. A module ${}_RF$ is called <u>a free R-module</u> if it is isomorphic to a direct sum of copies of ${}_RR$. A module ${}_RP$ is called <u>a projective R-module</u> if the following diagram can be completed, for every R-module ${}_RM$ and ${}_RN$ and every R-homomorphism f and g:

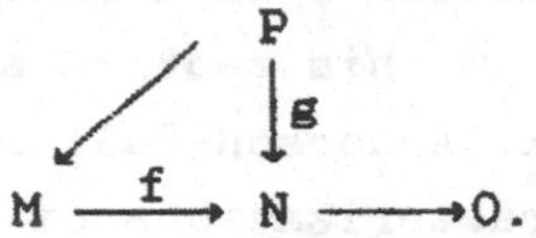

And a module ${}_RE$ is called <u>an injective R-module</u> if the following diagram can be completed, for every R-module ${}_RM$ and ${}_RN$ and every R-homomorphism f and g:

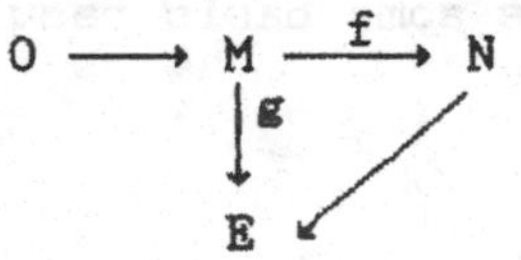

THEOREM 1.1. The following conditions are equivalent for a left R-module P:

(1) P is a projective R-module;

(2) $\mathrm{Hom}_R(P,-)$ is an exact functor, that is if

$$0 \longrightarrow M \longrightarrow N \longrightarrow L \longrightarrow 0$$

is an exact sequence of left R-modules, then

$$0 \longrightarrow \mathrm{Hom}_R(P,M) \longrightarrow \mathrm{Hom}_R(P,N) \longrightarrow \mathrm{Hom}_R(P,L) \longrightarrow 0$$

is an exact sequence of abelian groups;

(3) P is a direct summand of a free R-module;

(4) Every exact sequence $0 \longrightarrow M \longrightarrow N \longrightarrow P \longrightarrow 0$, where M and N are left R-modules, spilits.

THEOREM 1.2. The following conditions are equivalent for a left R-module E:

(1) E is an injective R-module;

(2) $\mathrm{Hom}_R(-,E)$ is an exact functor, that is if

$$0 \longrightarrow M \longrightarrow N \longrightarrow L \longrightarrow 0$$

is an exact sequence of left R-modules, then

$$0 \longrightarrow \mathrm{Hom}_R(L,E) \longrightarrow \mathrm{Hom}_R(N,E) \longrightarrow \mathrm{Hom}_R(M,E) \longrightarrow 0$$

is an exact sequence of abelian groups;

(3) E is a direct summand of every module of which it is a submodule;

(4) Every exact sequence $0 \longrightarrow E \longrightarrow N \longrightarrow L \longrightarrow 0$, where N and L are left R-modules, spilits;

(5) (Baer's lemma) For every left ideal I of R and every R-homomorphism g, the following diagram can be completed:

$$\begin{array}{ccccc} 0 & \longrightarrow & I & \stackrel{i}{\longrightarrow} & R \\ & & \downarrow g & \swarrow & \\ & & E & & \end{array}$$

where i is the inclusion map. In this case there is an $x \in E$ with $g(a) = ax$ for all $a \in I$.

An arbitrary direct sum of projective modules is a projective module. An arbitrary direct product of injective modules is an injective module. Every module can be embedded in an injective module.

Let M be a module and N a submodule. N is called an <u>essential</u> (or <u>large</u>) submodule of M, or M is called an <u>essential extension</u> of N, if for any nonzero submodule L of M we have $L \cap N \neq 0$. And N is called a <u>superfluous</u> (or <u>small</u>) submodule of M, if for any submodule $L \neq M$ we have $L + N \neq M$.

For every module M, there is an injective module $E(M)$ which is an essential extension of M. This $E(M)$ is unique up to isomorphism, so it is called <u>the injective envelope of M</u>. If N is an essential submodule of M, then $E(M) = E(N)$. If $M = N_1 \oplus N_2$ then $E(M) = E(N_1) \oplus E(N_2)$.

A projective module P is called <u>a projective cover of a module M</u>, if there is an epimorphism $f\colon P \longrightarrow M$ such that

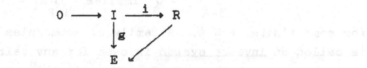

Ker(f) is a superfluous submodule of P.

While every module has an injective envelope, not every module has a projective cover. For example, the simple $\mathbb{Z}$-module $\mathbb{Z}/2\mathbb{Z}$ has no projective covers.

A module M is called <u>finitely generated</u> in case for every set $\{M_i\}_A$ of submodules of M that spans M, there is a finite subset $F \subseteq A$ such that $\{M_i\}_F$ spans M; that is,

$$\sum_{i\in A} M_i = M \quad \text{implies} \quad \sum_{i\in F} M_i = M$$

for some finite $F \subseteq A$. Dually, a module M is called <u>finitely cogenerated</u> in case for every set $\{M_i\}_A$ of submodules of M

$$\bigcap_{i\in A} M_i = 0 \quad \text{implies} \quad \bigcap_{i\in F} M_i = 0$$

for some finite $F \subseteq A$. A family of submodules $\{M_i\}_{i\in I}$ of M is called an <u>inverse system</u> in case for any pair $i, j \in I$, there is a $k \in I$ such that $M_k \subseteq M_i \cap M_j$. In this event it follows from induction that for any finite number of indices $i_1, \ldots, i_n \in I$, there is a $k \in I$ such that $M_k \subseteq M_{i_1} \cap \ldots \cap M_{i_n}$.

PROPOSITION 1.3. The following statements are equivalent for a module M:

(1) M is finitely generated;

(2) M contains a finite spanning set;

(3) For every indexed set $\{G_i\}_A$ of modules and epimorphisms $\bigoplus_{i\in A} G_i \longrightarrow M \longrightarrow 0$, there is a finite set $F \subseteq A$ and an epimorphisms $\bigoplus_{i\in F} G_i \longrightarrow M \longrightarrow 0$.

A module $T \neq 0$ is called <u>simple</u> in case T and 0 are the only submodules of T. A module M is called <u>semisimple</u> in case it is a direct sum of simple submodules.

PROPOSITION 1.4. The following are equivalent for a module M:

(1) M is semisimple;

(2) M is a sum of some set of simple submodules;

(3) M is a sum of its simple submodules;

(4) Every submodule of M is a direct summand;

(5) Every short exact sequence

$$0 \longrightarrow K \longrightarrow M \longrightarrow N \longrightarrow 0$$

splits.

PROPOSITION 1.5. The following statements are equivalent for a module M:

(1) M is finitely cogenerated;

(2) $E(M) \cong E(T_1) \oplus \ldots \oplus E(T_n)$ for some finitely many simple modules $T_1, \ldots, T_n$;

(3) For every indexed set $\{U_i\}_A$ of modules and monomorphisms $0 \longrightarrow M \longrightarrow \prod_{i \in A} U_i$, there is a finite set $F \subseteq A$ and a monomorphism $0 \longrightarrow M \longrightarrow \prod_{i \in F} U_i$;

(4) Every inverse system of nonzero submodules of M is bounded below by a nonzero submodule of M.

A module M is called an <u>artinian module</u> in case for every chain of submodules

$$M_1 \geqq M_2 \geqq \ldots \geqq M_n \geqq \ldots$$

there is an n with $M_{n+i} = M_n$ $(i=1,2,\ldots)$. An artinian module is said to satisfy the <u>descending chain condition</u>. Dually, a module M is called a <u>noetherian module</u> in case for every chain of submodules

$$M_1 \leqq M_2 \leqq \ldots \leqq M_n \leqq \ldots$$

there is an n with $M_{n+i} = M_n$ $(i=1,2,\ldots)$. A noetherian module is said to satisfy the <u>ascending chain condition</u>. A module M has <u>finite length</u> in case it is both artinian and noetherian. It is well-known that a module has finite length if and only if M has a <u>composition series</u>:

$$M = M_0 > M_1 > \dots > M_n = 0,$$

where each composition factor M_i/M_{i+1} is a simple module and n is a natural number. The Jordan-Holder Theorem states that if M has a second composition series:

$$M = N_0 > N_1 > \dots > N_m = 0,$$

then $n = m$ and there is a permutation p of $\{1,2,\dots,n\}$ such that

$$M_i/M_{i+1} \cong N_{p(i)}/N_{p(i+1)} \qquad (i = 1,2,\dots,n).$$

In this case, we denote $n = c(M)$. If M does not have finite length, we write $c(M) = \infty$. A ring R is called left (right) artinian in case the module ${}_RR$ (R_R) is an artinian module; and R is called left (right) noetherian in case the module ${}_RR$ (R_R) is a noetherian module.

PROPOSITION 1.6. The following statements are equivalent for a module M:

(1) M is artinian;

(2) Every non-empty set of submodules of M has a minimal element;

(3) Every factor module of M is finitely cogenerated.

PROPOSITION 1.7. The following statements are equivalent for a module M:

(1) M is noetherian;

(2) Every non-empty set of submodules of M has a maximal element;

(3) Every submodule of M is finitely generated.

PROPOSITION 1.8. Let $0 \longrightarrow L \longrightarrow M \longrightarrow N \longrightarrow 0$ be an exact sequence of modules. Then

(1) If M is finitely generated then so is N; and if both L and N are finitely generated then so is M.

(2) If M is finitely cogenerated then so is L; and if

both L and N are finitely cogenerated then so is M.

(3) M is artinian (noetherian) if and only if both L and N are artinian (noetherian).

Let M be a module. The radical of M, denoted by Rad(M), is the intersection of all maximal submodules of M. The socle of M, denoted by Soc(M), is the sum of all simple submodules of M. If R is a ring, then $Rad({}_RR) = Rad(R_R)$. So we denote $J(R) = Rad({}_RR)$, which is called the Jacobson radical of R. Nakayama's Lemma states that if ${}_RM$ is a finitely generated module with M = J(R)M then M = 0. For a ring R, the left module ${}_RR$ is semisimple if and only if the right module R_R is semisimple; in this case R is called a semisimple ring. A ring R is called simple in case the only (two-sided) ideals of R are 0 and R itself. A simple ring is left artinian if and only if it is right artinian; in this case R is called simple artinian. Every simple artinian ring is isomorpgic to $M_n(D)$, which is the ring of n by n matrices over some division ring D. The Wedderburn-Artin Theorem states that every semisimple ring is a finite direct sum of simple artinian rings. A ring R is called semilocal in case R/J(R) is semisimple. If R is a semilocal ring with Jacobson radical J, then for every module ${}_RM$, we have JM = Rad(M), M/JM is semisimple, and $Soc(M) = \{m \in M \mid Jm = 0\}$ which is denoted by $Ran_M(J)$.

PROPOSITION 1.9. Let M be a module. Then

(1) M is finitely generated if and only if M/Rad(M) is finitely generated and Rad(M) is superfluous in M; and

(2) M is finitely cogenerated if and only if Soc(M) is finitely generated and Soc(M) is essential in M.

Let ${}_RM$ be a left R-module. Then for each $X \subseteq M$, the (left) annihilator of X in R is

$$Lan_R(X) = \{r \in R \mid rX = 0\};$$

and, for each $I \subseteq R$, the (right) annihilator of I in M is

$$\mathrm{Ran}_M(I) = \{m \in M \mid Im = 0\}.$$

The module ${}_RM$ is called a <u>faithful module</u> in case $\mathrm{Lan}_R(M) = 0$. Beginning with a right R-module M_R, we encounter the right annihilator $\mathrm{Ran}_R(X)$ and the left annihilator $\mathrm{Lan}_M(I)$, and the right R-module M_R is faithful if $\mathrm{Ran}_R(M) = 0$.

PROPOSITION 1.10. If ${}_RE$ is an injective module, then for any ideal I of R the R/I-module $\mathrm{Ran}_E(I)$ is injective.

PROOF. Let $U = E({}_{R/I}\mathrm{Ran}_E(I))$, the injective envelope of the left R/I-module $\mathrm{Ran}_E(I)$. Then U is a canonical left R-module, and is essential over $\mathrm{Ran}_E(I)$. Therefore, U can be embedded in the injective envelope $E({}_R\mathrm{Ran}_E(I)) \subseteq {}_RE$, hence $\mathrm{Ran}_E(I) \subseteq U \subseteq E$. But $IU = 0$, so $U \subseteq \mathrm{Ran}_E(I)$, and $U = \mathrm{Ran}_E(I)$ as asserted.

PROPOSITION 1.11. If ${}_RE$ is a finitely cogenerated injective module and $S = \mathrm{End}({}_RE)$, then $J(S) = \mathrm{Ran}_S(\mathrm{Soc}({}_RE))$; and $S/J(S) \cong \mathrm{End}({}_R\mathrm{Soc}({}_RE))$.

Let R be a ring. Denote R-Mod (Mod-R) the category of left (right) R-modules. For a module ${}_RG$, let $G^{(A)}$ (G^A) denote the direct sum (product) of $|A|$-copies of G. Then G is called a <u>generator in R-Mod</u> in case for every module ${}_RM$, there is an epimorphism $G^{(A)} \longrightarrow M$ for some set A. Every generator is a faithful module. A module ${}_RE$ is called a <u>cogenerator in R-Mod</u> in case for every module ${}_RM$, there is a monomorphism $M \longrightarrow E^A$ for some set A. It is immediate that ${}_RE$ is a cogenerator only if $\mathrm{Hom}_R(M,E) \neq 0$ for all ${}_RM \neq 0$. Every cogenerator is a faithful module.

PROPOSITION 1.12. Let E be a cogenerator in R-Mod and let M be a module in R-Mod with a submodule N. If $m \in M\backslash N$, then there is an $f \in \mathrm{Hom}_R(M, E)$ such that $f(N) = 0$ but $f(m) \neq 0$.

PROOF. Let $h : M/N \longrightarrow E^A$ be a monomorphism for some set A. Since $m+N$ is a nonzero element in M/N, there is a projection $p : E^A \longrightarrow E$ such that $ph(m+N) \neq 0$. Then $f = phg \in \mathrm{Hom}_R(M,E)$ has the required property, where $g : M \longrightarrow M/N$ is the natural map.

PROPOSITION 1.13. (1) A projective module is a generator if each simple module is its epimorphic image; and

(2) An injective module is a cogenerator if it contains each simple module.

Let R and S be rings. R-Mod and S-Mod are called <u>equivalent</u>, in case there are covariant functors F: R-Mod $\longrightarrow$ S-Mod and G: S-Mod $\longrightarrow$ R-Mod such that $GF \cong 1_{R\text{-Mod}}$ and $FG \cong 1_{S\text{-Mod}}$, where 1 denotes the identity functors. In this case, the rings R and S are called (<u>Morita</u>) <u>equivalent</u>.

PROPOSITION 1.14. Let F: R-Mod $\longrightarrow$ S-Mod define an equivalence. Let ${}_RM$ be an R-module. Then

(1) $\mathrm{End}({}_RM) \cong \mathrm{End}({}_SFM)$;

(2) ${}_RM$ is projective (injective) if and only if ${}_SF(M)$ is projective (injective);

(3) ${}_RM$ is a generator (cogenerator) if and only if ${}_SF(M)$ is a generator (cogenerator);

(4) ${}_RM$ is finitely generated (finitely cogenerated) if and only if ${}_SF(M)$ is finitely generated (finitely cogenerated);

(5) ${}_RM$ is semisimple if and only if ${}_SF(M)$ is semisimple;

(6) ${}_RM$ is artinian (noetherian) if and only if ${}_SF(M)$ is artinian (noetherian), and $c({}_RM) = c({}_SF(M))$.

A finitely generated projective generator is called a <u>progenerator</u>.

THEOREM 1.15. Let R and S be rings and let $F: R\text{-Mod} \longrightarrow S\text{-Mod}$ and $G: S\text{-Mod} \longrightarrow R\text{-Mod}$ be additive functors. Then F and G are inverse equivalence if and only if there exists a bimodule ${}_RP_S$ such that

(1) ${}_RP$ and P_S are progenerators;

(2) ${}_RP_S$ is balanced;

(3) $F \cong \mathrm{Hom}_R(P,-)$ and $G \cong (P\otimes_S -)$.

Moreover, if there is a bimodule ${}_RP_S$ satisfying these conditions, then with $Q = \mathrm{Hom}_R(P,R)$, we have ${}_SQ_R$ with ${}_SQ$ and Q_R progenerators and $F \cong (Q\otimes_R -)$ and $G \cong \mathrm{Hom}_S(Q,-)$.

In the assumptions of Theorem 1.15, we say that ${}_RP_S$ (or ${}_SQ_R$) defines an equivalence, or that R and S are equivalent via the bimodule ${}_RP_S$ (or ${}_SQ_R$).

COROLLARY 1.16. The following assertions are equivalent:

(1) R and S are equivalent;

(2) Mod-R and Mod-S are equivalent;

(3) There is a progenerator P_R with $S \cong \mathrm{End}(P_R)$;

(4) There is a progenerator ${}_RQ$ with $S \cong \mathrm{End}({}_RQ)$.

COROLLARY 1.17. Let R be a ring and let $n > 0$ be a natural number. Then R and $\mathbb{M}_n(R)$ are equivalent rings, where $\mathbb{M}_n(R)$ denotes the ring of $n \times n$ marices over R.

A ring R with $J = J(R$ is called <u>semiperfect</u> in case R is semilocal and for every idempotent $g+J$ of R/J there is an idempotent e of R with $g+J = e+J$. In particular, R is semiperfect if R is semilocal and J is nil. A <u>local ring</u> is a ring with a unique maximal left ideal, so local rings are semiperfect. A set of idempotents $e_1,\dots,e_n$ of a ring is called a <u>complete set of idempotents</u> in case $1 = e_1+\dots+e_n$ and $e_ie_j = 0$ if $i \neq j$.

THEOREM 1.18. For a ring R the following statements are equivalent:

(1) R is semiperfect;

(2) R has a complete set $e_1,\dots,e_n$ of primitive idempotents with each e_iRe_i a local ring;

(3) Every simple left R-module has a projective cover;

(4) Every finitely generated left R-module has a projective cover.

PROPOSITION 1.19. The endomorphism ring of a finitely cogenerated injective module is a semiperfect ring.

Let R be a semiperfect ring with radical J. A set $e_1,\dots,e_n$ of primitive idempotents of R is called <u>basic</u> in case $Re_1/Je_1,\dots,Re_n/Je_n$ is an irredundant set of representatives of the simple left R-modules. (Every complete set of primitive idempotents contains a basic subset.) In this case $e_1R/e_1J,\dots,e_nR/e_nJ$ is an irredundant set of representatives of the simple right R-modules, and $Re_1,\dots,Re_n$ $(e_1R,\dots,e_nR)$ is an irredundant set of representatives of the indecomposable projective left (right) R-modules. If ${}_RP$ is a projective R-module then there exist sets $A_1,\dots,A_n$ (unique to within cardinality and possibly empty) such that ${}_RP \cong Re_1^{(A_1)} \oplus\dots\oplus Re_n^{(A_n)}$. Hence if ${}_RP$ is finitely generated projective then ${}_RP \cong Re_1^{k_1} \oplus\dots\oplus Re_n^{k_n}$ for some non-negative integers $k_1,\dots,k_n$. The idempotent $e_1+\dots+e_n$ is then called a <u>basic idempotent</u> of R. The ring R is called a <u>basic ring</u> in case 1 is a basic idempotent of R. If e is a baisc idempotent of R then eRe is a basic semiperfect ring and this ring is called a <u>basic ring for R</u>, and in this case R and eRe are equivalent rings (Corollary 1.15). Moreover, two semiperfect rings are Morita equivalent if and only if their basic rings are isomorphic.

Let R be a ring and let $\{T_i\}_A$ be an irredundent set of representatives of the simple modules in R-Mod; that is, every simple module is isomorphic to T_i for some $i \in A$, and T_i is not isomorphic to T_j if $i \neq j$ in A. Let $U = \bigoplus_{i \in A} E(T_i)$. Then U is a cogenerator in R-Mod, and U is isomorphic to a submodule of every cogenerator in R-Mod, so U is called a minimal cogenerator for R-Mod. (According to a recent result of Osofsky [91], minimal cogenerators need not be unique up to isomorphism.) And $E(U) = E(\bigoplus_{i \in A} T_i)$ is called the minimal injective cogenerator for R-Mod. (Minimal injective cogenerator is unique up to isomorphism.) Of course, $U \cong E(U)$ if there are only finitely many non-isomorphic simple modules. This is so if R is a semiperfect ring.

Let R and S be rings and ${}_RE_S$ a bimodule.
There are canonical ring homomorphisms

$$\lambda\colon R \longrightarrow \mathrm{End}(E_S) \quad \text{and} \quad \rho\colon S \longrightarrow \mathrm{End}({}_RE)$$

via

$$\lambda(r)\colon x \longmapsto rx \quad \text{and} \quad \rho(s)\colon x \longmapsto xs,$$

where $r \in R$, $x \in E$ and $s \in S$. If both λ and ρ are epimorphisms (isomorphisms) we say that ${}_RE_S$ is a balanced bimodule (faithfully balanced bimodule).

If ${}_RM$ (N_S) is a left R-module (right S-module), then $\mathrm{Hom}_R({}_RM, {}_RE_S)$ ($\mathrm{Hom}_S(N,E)$) is a right S-module (left R-module). So there are the pair of contravariant additive functors $\mathrm{Hom}_R(-, {}_RE_S)$: R-Mod $\longrightarrow$ Mod-S and $\mathrm{Hom}_S(-, {}_RE_S)$: Mod-S $\longrightarrow$ R-Mod. For brevity we write

$$(\ \ )^* = \mathrm{Hom}(-, {}_RE_S)$$

to denote either of these functors. For each M in R-Mod or Mod-S

$$[\sigma_M(m)](f) = f(m) \qquad (m \in M,\ \ f \in M^*)$$

defines the evaluation homomorphism

$$\sigma_M : M \longrightarrow M^{**}.$$

A module M is called <u>E-reflexive</u> (<u>E-torsionless</u>) in case σ_M is an isomorphism (monomorphism). We see from the definition of σ_M that $m \in \mathrm{Ker}(\sigma_M)$ if and only if $m \in \mathrm{Ker}(f)$ for all $f : M \longrightarrow E$. It follows that M is E-torsionless if and only if there is an embedding $M \longrightarrow E^A$ for some set A.

THEOREM 1.20. Let ${}_RE_S$ be a bimodule and M be a module in R-Mod or Mod-S. Then M^* is always E-torsionless, and

(1) If M is E-reflexive, then M^* is also E-reflexive.

(2) If $M \cong M_1 \oplus \ldots \oplus M_n$, then M is E-reflexive if and only if each of $M_1, \ldots, M_n$ is E-reflexive.

(3) ${}_RE$ (E_S) is a faithful module if and only if ${}_RR$ (S_S) is E-torsionless; and ${}_RE_S$ is a faithfully balanced bimodule if and only if ${}_RR$ and S_S are E-reflexive.

(4) If $M_1 \xrightarrow{f} M_2$ in R-Mod or Mod-S, then the diagram

$$\begin{array}{ccc} M_1 & \xrightarrow{f} & M_2 \\ \downarrow & & \downarrow \\ M_1^{**} & \xrightarrow{f^{**}} & M_2^{**} \end{array}$$

is commutative. Thus the evaluation homomorphisms induces natural transformations

$$\sigma : 1_{R\text{-Mod}} \longrightarrow ((\ \ )^*)^*$$

and

$$\sigma : 1_{\text{Mod-}S} \longrightarrow ((\ \ )^*)^*.$$

If P_R and ${}_RM$ are R-modules, one form the <u>tensor product</u> of M and N denoted by $P \otimes_R M$, which is an abelian group. Every element in $P \otimes_R M$ is a finite sum of the form $p \otimes m$ where $p \in P$, $m \in M$. In case ${}_SP_R$, then ${}_S(P \otimes_R M)$. A module P_R is <u>flat</u> in case any exact sequence in R-Mod

$$0 \longrightarrow L \longrightarrow M \longrightarrow N \longrightarrow 0$$

induces an exact sequence

$$0 \longrightarrow P\otimes_R L \longrightarrow P\otimes_R M \longrightarrow P\otimes_R N \longrightarrow 0$$

in the category of abelian groups. The fact is that any projective module is flat. In conclusion of this chapter, we mention the following

THEOREM 1.21. Let R be a ring and ${}_RM$ an R-module. Then

(1) There is a left R-isomorphism $R\otimes_R M \cong M$ via $r\otimes m \longmapsto rm$;

(2) There is a left R-isomorphism $\mathrm{Hom}_R(R,M) \cong M$ via $f \longmapsto f(1)$;

(3) Adjoint Isomorphism. For rings R and S, consider the situation ${}_SM$, ${}_RN_S$, ${}_RE$. Then there is an abelian group isomorphism

$$F : \mathrm{Hom}_S({}_SM, \mathrm{Hom}_R(N,E)) \cong \mathrm{Hom}_R(N\otimes M, E),$$

where $F(f)(n\otimes m) = f(m)(n)$ and $F^{-1}(g)(m)(n) = g(n\otimes m)$, $f \in \mathrm{Hom}_S({}_SM, \mathrm{Hom}_R(N,E))$, $g \in \mathrm{Hom}_R(N\otimes M, E)$, $n \in N$, $m \in M$.

A ring R is called <u>semiprimary</u> in case R is semilocal and $J(R)$ is nilpotent. Hence semiprimary rings are semiperfect. The well-known Hopkin's Theorem asserts that R is left artinian if and only if R is semiprimary and left noetherian. A module is <u>indecomposable</u> if it can not be written as a direct sum of two proper submodules. In view of the first result of the following proposition, one notes that an injective module over a semiprimary ring (e.g., a left or right artinian ring) is indecomposable if and only if it is the injective envelope of a simple module.

PROPOSITION 1.22. Let R be a semiprimary ring with $J = J(R)$. If M is a left R-module, then

(1) $\mathrm{Soc}(M) = \mathrm{Ran}_M(J)$ is an essential submodule of M; and

(2) $\mathrm{Rad}(M) = JM$ is a superfluous submodule of M.

Moreover, if R is left artinian then for M the following statements are equivalent:

(a) M is finitely generated;

(b) M is noetherian;

(c) M has finite length;

(d) M is artinian;

(e) M/JM is finitely generated.

SECTION 2. BASIC CHARACTERIZATIONS OF MORITA DUALITY

In this section, we summarize some basic results in Morita duality that can be found in Anderson-Fuller [74, Sections 23, 24]. Let R and S be two fixed rings.

If C and D are two subcategories of R-Mod and Mod-S, then there is a duality between C and D in case there are contravariant additive functors $F: C \longrightarrow D$ and $G: D \longrightarrow C$ such that $GF \cong 1_C$ and $FG \cong 1_D$.

THEOREM 2.1 (Morita). Let C and D be full subcategories of R-Mod and Mod-S such that ${}_RR \in C$ and $S_S \in D$, and such that every module in R-Mod (Mod-S) isomorphic to one in C (D) is in C (D). If there is a duality between C and D induced by $F: C \longrightarrow D$ and $G: D \longrightarrow C$, then there is a bimodule ${}_RE_S$ such that (1) ${}_RE \cong G(S_S)$ and $E_S \cong F({}_RR)$; (2) There are natural isomorphisms $F \cong Hom_R(-,E)$ and $G \cong Hom_S(-, E)$; and (3) All modules M in C and all modules N in D are E-reflexive.

Let C and D be full subcategories of R-Mod and Mod-S such that ${}_RR \in C$ and $S_S \in D$, and C and D are closed under submodules and factor modules. Then a Morita duality is a duality between C and D, and in this case R has a (left) Morita duality and S has a right Morita duality. Since Theorem 2.1, we have the following equivalent definition:

DEFINITION. A bimodule ${}_RE_S$ defines a <u>Morita duality</u> in case (1) ${}_RR$ and S_S are E-reflexive; and (2) Every submodule and every factor module of an E-reflexive module is E-reflexive.

LEMMA 2.2 (Osofsky [66]). If a bimodule ${}_RE_S$ defines a Morita duality, then no infinite direct sum of non-zero left R-modules is E-reflexive.

Unlike Morita equivalence we have the following result in duality.

PROPOSITION 2.3. For no rings R and S is there a duality between R-Mod and Mod-S.

PROOF. If there is a duality via F: R-Mod ⟶ Mod-S and G: Mod-S ⟶ R-Mod, then by Theorem 2.1 there is a bimodule ${}_RE_S$ such that all M in R-Mod and all N in Mod-S are E-reflexive modules. Hence ${}_RE_S$ defines a Morita duality. By Lemma 2.2 it is impossible that every left R-module is E-reflexive.

THEOREM 2.4. The following statements are equivalent for a bimodule ${}_RE_S$:

(1) ${}_RE_S$ defines a Morita duality;

(2) Every factor module of ${}_RR$, S_S, ${}_RE$ and E_S is E-reflexive;

(3) ${}_RE_S$ is a balanced bimodule such that ${}_RE$ and E_S are injective cogenerators.

Colby and Fuller [84] have proved that if ${}_RE_S$ is any balanced bimodule such that ${}_RE$ and E_S are cogenerators, then ${}_RE$ and E_S are injective so ${}_RE_S$ defines a Morita duality. This shows that the "injective" condition in Theorem 2.4(3) is redundant.

COROLLARY 2.5. If ${}_RE_S$ defines a Morita duality and I is an ideal of R, then ${}_{R/I}\mathrm{Ran}_E(I)_{S/\mathrm{Ran}_S\mathrm{Ran}_E(I)}$ defines a Morita duality.

PROOF. Since ${}_RE_S$ is a balanced bimodule such that ${}_RE$ and E_S are injective cogenerators, ${}_{R/I}\mathrm{Ran}_E(I)_{S/\mathrm{Ran}_S\mathrm{Ran}_E(I)}$ is a balanced bimodule such that ${}_{R/I}\mathrm{Ran}_E(I)$ and $\mathrm{Ran}_E(I)_{S/\mathrm{Ran}_S\mathrm{Ran}_E(I)}$ are injective cogenerators. The injectivity of ${}_{R/I}\mathrm{Ran}_E(I)$ follows from Proposition 1.10, and that of $\mathrm{Ran}_E(I)_{S/\mathrm{Ran}_S\mathrm{Ran}_E(I)}$ follows from a right version of Proposition 1.10 and the fact that $\mathrm{Ran}_E(I) = \mathrm{Lan}_E\mathrm{Ran}_S\mathrm{Ran}_E(I)$. Then using Proposition 1.13(2), one sees that $\mathrm{Ran}_E(I)$ is a cogenerator on each side.

Let ${}_RE_S$ be a bimodule and let ${}_RM$ be a left R-module. For each $A \leq M$ the (<u>right</u>) <u>annihilator</u> of A in M^* is

$$\mathrm{Ran}_{M^*}(A) = \{ g \in M^* \mid A \leq \mathrm{Ker}(g)\};$$

and for each $B \leq M^*$ the (<u>left</u>) <u>annihilator</u> of B in M is

$$\mathrm{Lan}_M(B) = \{ m \in M \mid g(m) = 0 \text{ for all } g \in B \} = \cap \{\mathrm{Ker}(g) \mid g \in B\}.$$

Then one easily verifies that for all submodules M_1, M_2, M_i ($i \in I$) of M: (1) $M_1 \leq M_2$ implies $\mathrm{Ran}_{M^*}(M_1) \geq \mathrm{Ran}_{M^*}(M_2)$; (2) $M_1 \leq \mathrm{Lan}_M\mathrm{Ran}_{M^*}(M_1)$; (3) $\mathrm{Ran}_{M^*}(M_1) = \mathrm{Ran}_{M^*}\mathrm{Lan}_M\mathrm{Ran}_{M^*}(M_1)$; (4) $\mathrm{Ran}_{M^*}(\Sigma_I M_i) = \cap_I \mathrm{Ran}_{M^*}(M_i)$; and (5) $\mathrm{Ran}_{M^*}(\cap_I M_i) \geq \Sigma_I \mathrm{Ran}_{M^*}(M_i)$. Let N_S be a right S-module. Then for $A \leq N$ and $B \leq N^*$, we have $\mathrm{Lan}_{N^*}(A) = \{ g \in N^* \mid A \leq \mathrm{Ker}(g)\}$; and $\mathrm{Ran}_N(B) = \cap \{\mathrm{Ker}(g) \mid g \in B\}$.

THEOREM 2.6. Let ${}_RE_S$ define a Morita duality and let ${}_RM$ and N_S be E-reflexive modules. Then

(1) $\mathrm{End}({}_R M) \cong \mathrm{End}(\mathrm{Hom}_R(M,E)_S)$;

(2) For each submodule K of M and each submodule L of M^*,

$$\mathrm{Lan}_M(\mathrm{Ran}_{M^*}(K)) = K \quad\text{and}\quad \mathrm{Ran}_{M^*}(\mathrm{Lan}_M(L)) = L;$$

(3) For each submodule L of N and each submodule K of N^*,

$$\mathrm{Ran}_N(\mathrm{Lan}_{N^*}(L)) = L \quad\text{and}\quad \mathrm{Lan}_{N^*}(\mathrm{Ran}_N(K)) = K;$$

(4) For each index set $\{M_i\}_{i\in I}$ of submodules of M,

$$\mathrm{Ran}_{M^*}(\cap_I M_i) = \Sigma_I \mathrm{Ran}_{M^*}(M_i);$$

(5) For each index set $\{N_i\}_{i\in I}$ of submodules of N,

$$\mathrm{Lan}_{N^*}(\cap_I N_i) = \Sigma_I \mathrm{Lan}_{N^*}(N_i);$$

(6) The lattices of submodules of M and M^* are anti-isomorphic via the mapping $K \longmapsto \mathrm{Ran}_{M^*}(K)$;

(7) The lattices of submodules of N and N^* are anti-isomorphic via the mapping $L \longmapsto \mathrm{Lan}_{N^*}(L)$;

(8) The lattices of ideals of R and S are isomorphic, and the centers of R and S are isomorphic rings;

(9) Every finitely generated or finitely cogenerated left R- (right S-) module is E-reflexive;

(10) A left R- (right S-) module M is finitely generated projective if and only if M^* is finitely cogenerated injective;

(11) M is simple, semisimple, of finite length n, indecomposable, respectively, if and only if M^* is;

(12) M is noetherian if and only if M^* is artinian;

(13) M is finitely generated if and only if M^* is finitely cogenerated.

PROOF. The proofs of (1)-(3) and (6)-(10) are in Anderson -Fuller [74].

For (4): $\mathrm{Ran}_{M^*}(\cap_I M_i) \overset{(2)}{=} \mathrm{Ran}_{M^*}(\cap_I \mathrm{Lan}_M \mathrm{Ran}_{M^*}(M_i))$

$$= \mathrm{Ran}_{M^*}\mathrm{Lan}_M(\Sigma_I \mathrm{Ran}_{M^*}(M_i)) \overset{(2)}{=} \Sigma_I \mathrm{Ran}_{M^*}(M_i).$$

The proof of (5) is similar.

For (11): If M is simple or finite length n, then so is M^* by (6); if M is semisimple, then by Lemma 2.2, $M = M_1 \oplus \ldots \oplus M_n$ for finitely many simples $M_1, \ldots, M_n$ and so $M^* \cong (M_1)^* \oplus \ldots \oplus (M_n)^*$ is semisimple. The converse is similar since $M \cong M^{**}$. Since $M \cong M^{**}$ and $(M_1 \oplus M_2)^* \cong (M_1)^* \oplus (M_2)^*$, it follows that M is decomposable if and only if M^* is decomposable.

(12) also follows from the lattice anti-isomorphism of (6).

For (13): If there is a natural number n with an exact sequence $R^n \longrightarrow M \longrightarrow 0$, then there is an exact sequence $0 \longrightarrow M^* \longrightarrow (R^n)^*$. By (7) $(R^n)^*$ is finitely cogenerated, and so is M^*. For the converse, using (10) and Theorem 2.4, we know that ${}_RE$ and E_S are finitely cogenerated injective cogenerators. So if M^* is finitely cogenerated, then there is an exact sequenec $0 \longrightarrow M^* \longrightarrow (E_S)^n$ for some natural number n. This sequence induces an exact sequence in R-Mod: $((E_S)^n)^* \longrightarrow M^{**} \longrightarrow 0$. Now $((E_S)^n)^* \cong ({}_RR)^n$ and $M^{**} \cong M$, so M is finitely generated.

In view of Theorem 2.6(10), it is natural to ask whether or not the conditions "finitely generated" and "finitely cogenerated" can be dropped. We shall discuss this question at the end of Section 10.

THEOREM 2.7. If ${}_RE_S$ defines a Morita duality, then (1) ${}_RE_S$ is faithfully balanced; (2) both ${}_RE$ and E_S are finitely cogenerated injective cogenerators; and (3) Both R and S are semiperfect rings.

PROOF. Since ${}_RR$ and S_S are E-reflexive, (1) follows from Theorem 1.14; Since ${}_RE \cong (S_S)^*$ and $E_S \cong ({}_RR)^*$, so (2) follows from Theorems 2.6 and 2.4; Finally since $R \cong \mathrm{End}(E_S)$ and $S \cong \mathrm{End}({}_RE)$ by (1), hence R and S are semiperfect by (2) and

Proposition 1.19.

THEOREM 2.8. The following statements are equivalent:

(1) There exists a duality between the category R-FMod of finitely generated left R-modules and the category FMod-S of finitely generated right S-modules;

(2) R is left artinian and some bimodule ${}_RE_S$ defines a Morita duality;

(3) S is right artinian and some bimodule ${}_RE_S$ defines a Morita duality.

Moreover, if R, S and E satisfies either of the last two conditions, then a left R- (right S-) module is E-reflexive if and only if it is finitely generated if and only if it is finitely cogenerated.

REMARK 2.9. Even an (two-sided) artinian ring need not have a Morita duality. The counterexamples are formal triangular matrix rings.

If ${}_RM_S$ is a bimodule, we use $\begin{bmatrix} R & M \\ 0 & S \end{bmatrix}$ to denote the set of all $\begin{bmatrix} r & m \\ 0 & s \end{bmatrix}$, where $r \in R$, $m \in M$, $s \in S$. If one defines operations on $\begin{bmatrix} R & M \\ 0 & S \end{bmatrix}$ according to the usual rules for addition and multiplication of matrices:

$$\begin{bmatrix} r & m \\ 0 & s \end{bmatrix} + \begin{bmatrix} r' & m' \\ 0 & s' \end{bmatrix} = \begin{bmatrix} r+r' & m+m' \\ 0 & s+s' \end{bmatrix}$$

$$\begin{bmatrix} r & m \\ 0 & s \end{bmatrix} \begin{bmatrix} r' & m' \\ 0 & s' \end{bmatrix} = \begin{bmatrix} rr' & rm'+ms' \\ 0 & ss' \end{bmatrix}$$

then $\begin{bmatrix} R & M \\ 0 & S \end{bmatrix}$ becomes a ring, and it is called the formal triangular matrix ring constructed from R, M, S. It is easy to see that $\begin{bmatrix} R & M \\ 0 & S \end{bmatrix}$ is a left (right) artinian ring if and only if both R and S are left (right) artinian rings and ${}_RM$ (M_S) is a finitely generated module.

The following question was first proposed by E. Artin: Let C be a division subring of a division ring D. Is

$\dim({}_C D) = \dim(D_C)$? P.M. Cohn [66] gave a negative answer and showed that there is a division D and a division subring C of D such that $\dim(D_C)$ is finite but $\dim({}_C D)$ is not. Then $R = \begin{bmatrix} D & D \\ 0 & C \end{bmatrix}$ is an artinian ring without a Morita duality. We need a lemma before we prove this claim.

LEMMA 2.10. Let R be a semilocal ring with Jacobson radical J. If $J^2 = 0$ and ${}_R T$ is a simple module, then $E(T)/T \cong \mathrm{Hom}_R(J,T)$ as left R-modules.

PROOF. Let $E = E(T)$. The exact sequence of R-bimodules

$$0 \longrightarrow J \longrightarrow R \longrightarrow R/J \longrightarrow 0$$

induces an exact sequence of left R-modules

$$0 \longrightarrow \mathrm{Hom}_R(R/J,E) \longrightarrow \mathrm{Hom}_R(R,E) \longrightarrow \mathrm{Hom}_R(J,E) \longrightarrow 0.$$

Since $J^2 = 0$, ${}_R J = \mathrm{Ran}_J(J)$ is semisimple, so $\mathrm{Hom}_R(J,E) = \mathrm{Hom}_R(J,T)$ since $\mathrm{Soc}(E) = T$. The middle term $\mathrm{Hom}_R(R,E) \cong E$. Finally, $\mathrm{Hom}_R(R/J,E) \cong \mathrm{Ran}_E(J) = \mathrm{Soc}(E) = T$. And the result follows.

Now we show that the artinian ring $R = \begin{bmatrix} D & D \\ 0 & C \end{bmatrix}$ does not possess a Morita duality.

PROOF. Clearly $J(R) = \begin{bmatrix} 0 & D \\ 0 & 0 \end{bmatrix} = R\begin{bmatrix} 0 & 1 \\ 0 & 0 \end{bmatrix}$, so $J(R)^2 = 0$. Let $T = \begin{bmatrix} D & 0 \\ 0 & 0 \end{bmatrix}$ which is a simple left R-module, and then by the above lemma, $E(T)/T \cong \mathrm{Hom}_R(J,T)$. Now R is an artinian ring, hence by Theorem 2.8, R does not have a Morata duality if we show that the finitely cogenerated module $E(T)$ is not finitely generated. To do so, it suffices to prove that ${}_R H = \mathrm{Hom}_R(J,T)$ is not finitely generated. For any finitely many elements $h_i \in H$ $(i = 1,\dots,n)$, we shall show that $H \neq \Sigma_i Rh_i$. Suppose that $r_i = \begin{bmatrix} * & * \\ 0 & c_i \end{bmatrix} \in R$ $(i = 1,\dots,n)$, and let

$$h_i : \begin{bmatrix} 0 & 1 \\ 0 & 0 \end{bmatrix} \longmapsto \begin{bmatrix} d_i & 0 \\ 0 & 0 \end{bmatrix},$$

then $\Sigma r_i h_i : \begin{bmatrix} 0 & 1 \\ 0 & 0 \end{bmatrix} \longmapsto \begin{bmatrix} \Sigma c_i d_i & 0 \\ 0 & 0 \end{bmatrix}$. Since n is finite and $_C D$ is infinite dimensional, there exists some $d \in D$ with $d \neq \Sigma c_i d_i$. Define

$$h \in {}_R H = \mathrm{Hom}_R(J,T)$$

via

$$h : \begin{bmatrix} 0 & 1 \\ 0 & 0 \end{bmatrix} \longmapsto \begin{bmatrix} d & 0 \\ 0 & 0 \end{bmatrix}.$$

Then $h \neq \Sigma r_i h_i$. So $h \notin \Sigma R h_i$ and $H \neq \Sigma R h_i$.

SECTION 3. LINEARLY COMPACT MODULES

The notion of a linearly compact module plays an important role in Morita duality, as can be seen in the next section. This section contains some basic results on linearly compact modules.

Let M be a module. A family $\{m_i, M_i\}_{i \in I}$ (where $m_i \in M$ and $M_i \leq M$, $i \in I$) is called <u>solvable</u> in case there is an $m \in M$ such that $m - m_i \in M_i$ for all $i \in I$, it is called <u>finitely solvable</u> if $\{m_i, M_i\}_{i \in F}$ is solvable for any finite subset $F \subseteq I$, and the module M is called <u>linearly compact</u> in case any finitely solvable family of M is solvable. One notes that a family $\{m_i, M_i\}_{i \in I}$ is solvable if and only if $\bigcap_{i \in I} (m_i + M_i)$ is non-empty. Then a module M is linearly compact if and only if for any family $\{m_i, M_i\}_{i \in I}$, $\bigcap_{i \in I} (m_i + M_i)$ is non-empty if $\bigcap_{i \in F} (m_i + M_i)$ is non-empty for all finite subset $F \subseteq I$. A ring R is called <u>left (right) linearly compact</u> if the module $_R R$ (R_R) is linearly compact. The first four results of this section were essentially proved in Zelinsky [53].

LEMMA 3.1. Every artinian module is linearly compact.

PROOF. Let M be an artinian module and let $\{m_i, M_i\}_{i\in I}$ be a finitely solvable family of M. Consider the set

$$W = \{ \bigcap_{i\in F} M_i \mid F \text{ is a finite subset of } I \}.$$

Since M is artinian, W has a minimal element, say, $\bigcap_{i\in F} M_i$ for some finite subset $F \subseteq I$. Since $\{m_i, M_i\}_{i\in F}$ is solvable, there is some $m \in M$ with $m-m_i \in M_i$ for all $i \in F$. We shall show that $m-m_j \in M_j$ for all $j \in I$. Let $j \in I$. Since $F\cup\{j\}$ is still a finite subset of I, $\{m_i, M_i\}_{i\in F\cup\{j\}}$ is solvable. So there exists some $m' \in M$ such that $m'-m_i \in M_i$ for all $i \in F\cup\{j\}$. Take $i_0 \in F$, then $m-m' = (m-m_{i_0})-(m'-m_{i_0}) \in M_i$ for all $i \in F$. Then $m-m' \in \bigcap_{i\in F} M_i = (\bigcap_{i\in F} M_i)\cap M_j$, where the equality holds from the minimality of $\bigcap_{i\in F} M_i$ in W. It follows that $m-m' \in M_j$ and so $m-m_j = (m-m')+(m'-m_j) \in M_j$.

COROLLARY 3.2. A left (right) artinian ring is left (right) linearly compact.

PROPOSITION 3.3. Let $0 \longrightarrow L \longrightarrow M \longrightarrow N \longrightarrow 0$ be an exact sequence of modules. Then M is linearly compact if and only if both L and N are linearly compact.

PROOF. (==>) This is obvious from the definition.

(<==) Let L be a submodule of M and let $N = M/L$. Let $\{m_i, M_i\}_{i\in I}$ be a finitely solvable family of M. For any finite subset $F \subseteq I$, let $m_F \in M$ with $m_F-m_i \in M_i$ for all $i \in F$. Let $K = \{ F \subseteq I \mid F \text{ is finite}\}$. Then $\{m_i, M_i\}_{i\in I}\cup\{m_F, \bigcap_{i\in F} M_i\}_{F\in K}$ is still a finitely solvable family in M. Denote $\bigcap_{i\in F} M_i = M_F$. Now M/L is linearly compact and $(m_j+L, (M_j+L)/L)_{j\in I\cup K}$ is finitely solvable, so there is an element $m+L \in M/L$ such that

$m-m_j \in M_j+L$ for all $j \in I \cup K$. Hence $L \cap (m_j-m+M_j)$ is non-empty for all $j \in I \cup K$. Let $l_j = m_j-m+m_j' \in L \cap (m_j-m+M_j)$, where $m_j' \in M_j$, then $L \cap (m_j-m+M_j) = l_j + (M_j \cap L)$. If $F \in K$, we have $\bigcap_{i \in F} (l_i+(M_i \cap L)) = \bigcap_{i \in F} (m_i-m+M_i) \cap L = (m_F-m+M_F) \cap L$ is nonempty since $F \in K$. It follows that $\{l_i, M_i \cap L\}_{i \in I}$ is a finitely solvable family of the linearly compact module L. Hence there is some $l \in L$ with $l-l_i \in M_i \cap L$ for all $i \in I$. Then $m+l \in M$ and $(m+l)-m_i = (l-l_i)+l_i+m-m_i = (l-l_i)+m_i' \in M_i$ for all $i \in I$.

PROPOSITION 3.4. A linearly compact module can not be an infinite direct sum of nonzero submodules.

PROOF. Suppose that M is a linearly compact module and $M = \bigoplus_{i \in I} M_i$ is an infinite direct sum of nonzero submodules M_i's. For each $i \in I$, let $M_{(i)} = \bigoplus_{j \neq i} M_j$ and let $0 \neq m_i \in M_i$. For any finite subset $F \subseteq I$, we have $(\sum_{j \in F} m_j)-m_i \in M_{(i)}$ for all $i \in F$. So $\{m_i, M_{(i)}\}_{i \in I}$ is a finitely solvable family of the linearly compact module M. It follows that there is some $m \in M$ such that $m-m_i \in M_{(i)}$ for all $i \in I$. Hence for each $i \in I$, the i-component of m is m_i $(\neq 0)$. This is a contradiction.

Using a similar method, one shows that for any ring R, the polynomial ring $R[x]$ is not left linearly compact.

COROLLARY 3.5. If R is a left artinian ring, then every linearly compact R-module ${}_RM$ is finitely generated (i.e., has finite length).

PROOF. Let $J = J(R)$. By Proposition 3.3, M/JM is still linearly compact. Being semisimple, M/JM is finitely generated. Hence M is finitely generated by Corollary 1.22.

If R is a ring, we form the formal power series ring $R[[x]]$ over R. The elements in $R[[x]]$ are formal power series $\Sigma_{i\geq 0} r_i x^i$, with $r_i \in R$ for all $i \geq 0$. The addition of $R[[x]]$ is given componently, and the multiplication is given by

$$(\Sigma_{i\geq 0} r_i x^i)(\Sigma_{i\geq 0} s_j x^j) = \Sigma_{i\geq 0} t_k x^k$$

where $t_k = \Sigma_{i+j=k} r_i s_j$, for any $r_i, s_j \in R$. Unlike polynomial rings, we have

PROPOSITION 3.6. If F is a field then $F[[x]]$ is linearly compact.

PROOF. Each proper ideal of $R = F[[x]]$ is of the form Rx^i for some $i > 0$. To show a finitely solvable family $\{m_i, M_i\}_{i\in I}$ is solvable, we can assume that each $M_i \neq 0$ and $M_i \neq M$. So let $\{f_i(x), Rx^{k_i}\}_{i\in \mathbb{N}}$ be a finitely solvable family, where $k_1 < k_2 < k_3 < \ldots$. Let $f_i(x) = \Sigma_{i\geq 0} a_{ij} x^j$. For any natural number n, $F = \{k_1, k_2, \ldots, k_n\}$ is a finite subset of $\mathbb{N}$, so there is some $f_F(x) = \Sigma_{i\geq 0} a_j x^j \in R$ such that $f_F(x)-f_i(x) \in Rx^{k_i}$ for all $k_i \in F$. Since $Rx^{k_n} \subseteq Rx^{k_i}$ for $k_i \in F$,

$$f_i(x) - f_n(x) = (f_F(x)-f_n(x)) - (f_F(x)-f_i(x)) \in Rx^{k_i}.$$

It follows that $a_{ij} = a_{nj}$ for all $j < k_i$ and $i = 1,\ldots,n$. Since n is arbitrary, we have: $a_{1j} = a_{2j} = \ldots$ for $j < k_1$; $a_{k_1 j} = a_{(k_1+1)j} = \ldots$ for $j < k_2$; .... Set $f(x) =$

$$a_{10} + \ldots + a_{1(k_1-1)} x^{k_1-1} + a_{k_1 k_1} x^{k_1} + \ldots + a_{k_1(k_2-1)} x^{k_2-1} + a_{k_2 k_2} x^{k_2} + \ldots.$$

Then we see that $f(x)-f_i(x) \in Rx^{k_i}$ for all i.

QUESTION 3.7. If R is a left linearly compact ring, is $R[[x]]$ left linearly compact?

THEOREM 3.8 (Leptin [55]). Let $f : {}_RM \longrightarrow {}_RX$ be an R-module homomorphism and let ${}_RM$ be linearly compact. If $\{M_i\}_{i\in I}$ is an inverse system of ${}_RM$ (recall the definition from Section 1), then $f(\bigcap_{i\in I} M_i) = \bigcap_{i\in I} f(M_i)$.

PROOF. Let $x = f(m_i) \in \bigcap_{i\in I} f(M_i)$, where $m_i \in M_i$ for all i. Then $m_i - m_j \in Ker(f)$ for all i, j. If F is a finite subset of I, there is an $k \in I$ such that $M_k \subseteq \bigcap_{i\in F} M_i$. Then $m_k - m_i \in M_i$ for all $i \in F$. This shows that $\{m_i, M_i \cap Ker(f)\}_{i\in I}$ is a finitely solvable family of ${}_RM$, so there exists a $m \in M$ such that $m - m_i \in M_i \cap Ker(f)$ for all $i \in I$. Then $m \in \bigcap_{i\in I} M_i$ and $x = f(m_i) = f(m) \in f(\bigcap_{i\in I} M_i)$.

COROLLARY 3.9. If M is a linearly compact module with an inverse system $\{M_i\}_{i\in I}$, then for any submodule N of M it holds that $\bigcap_{i\in I} (N + M_i) = N + (\bigcap_{i\in I} M_i)$.

PROOF. Applying the above theorem to the natural homomorphism $M \longrightarrow M/N$.

COROLLARY 3.10. Let M be a linearly compact module with submodules $\{M_i\}_{i\in I}$. If N is a submodule of M such that M/N is finitely cogenerated and $\bigcap_{i\in I} M_i \subseteq N$ then there is some finite subset $F \subseteq I$ such that $\bigcap_{i\in F} M_i \subseteq N$.

PROOF. Since $W = \{ \bigcap_{i\in F} M_i \mid F \text{ is a finite subset of } I\}$ is an inverse system of M, by Corollary 3.9,

$$\bigcap_{M'\in W} (N + M') = N + (\bigcap_{M'\in W} M') = N + (\bigcap_{i\in I} M_i) = N.$$

Hence $N + M' = N$ for some $M' \in W$ by Proposition 1.5, since M/N is finitely cogenerated.

To see that a left linearly compact ring is semiperfect, we introduce the following concept. A module M is called complemented in case for any $N \leq M$, there is a $L \leq M$ such that L is minimal with respect to the property that $N + L = M$. In this case, L is called a complement of N in M. The following result was proved in Sandomierski [72].

PROPOSITION 3.11. If M is a linearly compact module, then M is complemented.

PROOF. Let $N \leq M$ and $W = \{ L \mid L \leq M \text{ and } N + L = M \}$. Note that N is also linearly compact by Proposition 3.3. Partially order W by $L_1, L_2 \in W$, $L_1 \leqq L_2$ if and only if $L_1 \supseteq L_2$. Let $\{L_i\}_{i\in I}$ be a totally ordered subset of W. Take $m \in M$, then $m = n_i + l_i$ for some $n_i \in N$ and $l_i \in L_i$. If F is a finite subset of I, then there is a $j \in F$ with $L_j \subseteq L_i$ for all $i \in F$ hence $n_j - n_i = l_i - l_j \in L_i$ for each $i \in F$. This means $\{n_i, N \cap L_i\}_{i\in I}$ is a finitely solvable family of the linearly compact module N. Let $n \in N$ with $n - n_i \in N \cap L_i$ for all $i \in I$. Then $m - n = m - n_i + n_i - n = l_i + n_i - n \in L_i$ for all $i \in I$. So $m - n \in \bigcap_{i\in I} L_i$. Since m is an arbitrary element in M, we have $M = N + (\bigcap_{i\in I} L_i)$ and $(\bigcap_{i\in I} L_i)$ is an upper bound for $\{L_i\}_{i\in I}$ in W. So by Zorn's Lemma W has a maximal element L in W, and L will be a complement of N in M.

LEMMA 3.12. If $M = N + L$ where L is a complement of N in M, then $L \cap N$ is superfluous in L.

PROOF. If $(L \cap N) + K = L$ for some $K \leq L$, then $M = N + L = N + (L \cap N) + K = N + K$. It follows that $K = L$.

There is a more general result than Proposition 3.13 in Kasch's book [82, Theorem 11.1.5].

PROPOSITION 3.13. A ring R is semiperfect if and only if the module ${}_RR$ is complemented.

PROOF. (<==) Let N be a left ideal of R. We show that the cyclic R-module R/N has a projective cover. Then R is semiperfect by Theorem 1.17(3).

Let ${}_RR = N + L$ where L is a complement of N in ${}_RR$. Let $n : {}_RR \longrightarrow R/N$ be the natural map. Then $n|_L : L \longrightarrow R/N$ is an epimorphism with $\ker(n|_L) = L \cap N$ that is superfluous in L by Lemma 3.12. To show that L is a projective cover of R/N, it suffices to establish the projectivity of L. We shall prove that L is, in fact, a summand of ${}_RR$.

Since L is a left ideal, it has a complement L_1. Let

$$g : R = L_1 + L \longrightarrow R/(L_1 \cap L)$$

be the natural map, then $g(R) = g(L_1) \oplus g(L)$. If $t : g(R) \longrightarrow g(L)$ is the projection, then a commutative diagram exists

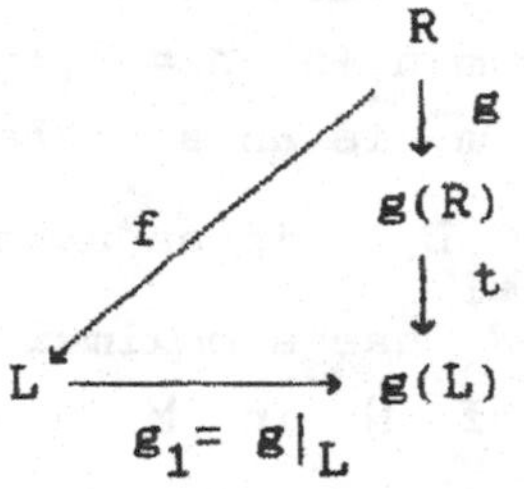

As $tg = g_1f$, $g_1(L) = g(L) = tg(L) = g_1f(L)$, thus $L = f(L) + \ker(g_1)$. By Lemma 3.12, $\ker(g_1) = L \cap L_1$ is superfluous in L, then $L = f(L)$. Hence $R = L + \ker(f)$. As $\ker(f) \leqq \ker(tg) = L_1$ and from the minimality of L_1 it follows that $\ker(f) = L_1$. On the other hand we have

$$L_1 = \ker(tg) = \ker(g_1 f) = f^{-1}(\ker(g_1)) = f^{-1}(L \cap L_1),$$

and since f is an epimorphism, we get
$0 = f(L_1) = ff^{-1}(L \cap L_1) = L \cap L_1$. Therefore $R = L \oplus L_1$ as promised.

($\Longrightarrow$) Let N be a left ideal of R and $n : R \longrightarrow R/N$ the natural map. Since R/N has a projective cover, it is routine to verify (or see Anderson and Fuller [74, Lemma 17.17]) that there is a decomposition ${}_RR = P \oplus Q$, such that $f = n|_P : P \longrightarrow R/N$ is a projective cover. So ${}_RR = P + N$. We shall prove that P is a complement of N in ${}_RR$. To do so, let $M \leq P$ and ${}_RR = M + N$, then $f(P) = n(P) = n(R) = n(M) = f(M)$. So $P = f^{-1}f(M) = M + \ker(f)$. Since $\ker(f)$ is superfluous in P, it follows that $M = P$.

COROLLARY 3.14. If a ring R is left linearly compact, then R is semiperfect.

In conclusion of this section, we shall see that linearly compact modules are preserved by equivalence. These results are due to Sandomierski [72a, 72b].

LEMMA 3.15. Let ${}_RP$ be a finitely generated projective R-module and $S = \mathrm{End}({}_RP)$. If ${}_RM$ is an R-module and ${}_SN \leq {}_S\mathrm{Hom}_R(P,M)$, then $N = \mathrm{Hom}_R(P,NP)$, where NP is the submodule of ${}_RM$ generated by $\{\mathrm{Im}(g) \mid g \in N\}$.

PROOF. Clearly $N \subseteq \mathrm{Hom}_R(P,NP)$. Let $g \in \mathrm{Hom}_R(P,NP)$, then let $j_n : P^{(N)} \longrightarrow P$ be the n-th projection where $n \in N$. Now the following diagram commutative with exact row

$$\begin{array}{ccccc} & & P & & \\ & \swarrow^{g'} & \downarrow g & & \\ P^{(N)} & \xrightarrow{\ f\ } & NP & \longrightarrow & 0 \end{array}$$

where $f(\{p_n\}_{n\in N}) = \sum_{n\in N} n(p_n)$ and g' exists since $_RP$ is projective. Since $_RP$ is finitely generated there is a finite subset F of N such that $j_n g'(p) = 0$ for all $p \in P$ and all $n \notin F$. Now

$$g(p) = fg'(p) = \sum_{n\in F} n(j_n g'(p))$$

hence $g = \sum_{n\in F} nj_n g'$. Since $j_n g' \in S$, $g \in N$, and the lemma follows.

THEOREM 3.16. Let $_RP$ be a finitely generated projective R-module and $S = \mathrm{End}(_RP)$. If $_RM$ is a linearly compact R-module, then $_S\mathrm{Hom}_R(P,M)$ is a linearly compact S-module.

PROOF. Let $\{f_i, N_i\}_{i\in I}$ be a finitely solvable family of $_S\mathrm{Hom}_R(P,M)$. For a fixed $p \in P$, consider the family $\{f_i(p), N_iP\}_{i\in I}$ in $_RM$, where N_iP is the submodule of $_RM$ generated by $\{\mathrm{Im}(g) \mid g \in N_i\}$. If F is a finite subset of I, there is an $f_F \in {_S\mathrm{Hom}_R(P,M)}$ such that $f_F - f_i \in N_i$ for each $i \in F$, hence $f_F(p) - f_i(p) \in N_iP$. Since $_RM$ is linearly compact, there is an $m_p \in M$ with $m_p - f_i(p) \in N_iP$ for all $i \in I$.

Let $j_i : M \longrightarrow M/N_iP$ be the natural map and let $h = \Pi j_i : M \longrightarrow \prod_{i\in I} (M/N_iP)$. Define a map $g : P \longrightarrow \mathrm{Im}(h)$ via $g(p) = h(m_p)$. Clearly, g is well-defined, since if m_p' is a solution of the family $\{f_i(p), N_iP\}_{i\in I}$, then $m_p - m_p' = (m_p - f_i(p)) - (m_p' - f_i(p)) \in N_iP$ for each $i \in I$, hence $h(m_p) = h(m_p')$. It is evident that g is an R-module homomorphism, and since $_RP$ is projective there is an $f \in \mathrm{Hom}_R(P,M)$ such that $g = hf$. Now for each $p \in P$ and each $i \in I$, $f(p) - f_i(p) \in N_iP$, since $h(f(p) - f_i(p)) = g(p) - hf_i(p) = h(m_p - f_i(p))$ and $m_p - f_i(p) \in N_iP$. It follows that $f - f_i \in \mathrm{Hom}_R(P, N_iP) = N_i$ by the above lemma. Hence $\{f_i, N_i\}_{i\in I}$ is solvable, and so $_S\mathrm{Hom}_R(P,M)$ is linearly compact.

COROLLARY 3.17. Let R and S be equivalent rings via the bimodule ${}_RP_S$. If ${}_RM$ be an R-module, then ${}_RM$ is linearly compact if and only if ${}_S\mathrm{Hom}_R(P,M)$ is linearly compact.

COROLLARY 3.18. If R is a left linearly compact ring and ${}_RP$ is a finitely generated projective module, then $S = \mathrm{End}({}_RP)$ is a left linearly compact ring.

PROOF. Since ${}_RR$ is linearly compact and ${}_RP$ is finitely generated, ${}_RP$ is also linearly compact. Hence by Theorem 3.16, ${}_SS = \mathrm{Hom}_R(P,P)$ is linearly compact.

SECTION 4. MORITA DUALITY AND LINEAR COMPACTNESS

Müller [70] has demonstrated the connections between linearly compact modules and Morita duality. Most results in this section are due to Müller [70], and other contributors are Ánh [82], Onodera [72], Sandomierski [72], and Vámos [77].

Let E and N be modules. Then E is called a <u>N-injective module</u> in case for any submodule L of N, any homomorphism $L \longrightarrow E$ can be entended to $N \longrightarrow E$.

From now on, "a duality" will mean "a (left) Morita duality" and "a right duality" will mean "a right Morita duality".

THEOREM 4.1. Let ${}_RE$ be a cogenerator and $S = \mathrm{End}({}_RE)$. Let ${}_RM$ be an R-module. The following statements are equivalent:

(1) ${}_RM$ is linearly compact;

(2) ${}_RM$ is E-reflexive and E_S is an M^*-injective right S-module.

PROOF. (1)==>(2). Let $0 \neq m \in M$. Since ${}_RE$ is a cogenerator, by Proposition 1.12 there is an $f \in M^*$ with

$f(m) \neq 0$. Hence $\sigma_M(m) \neq 0$, and σ_M is monic.

To see that σ_M is epic, let $h \in M^{**}$. If $f_1,\ldots,f_n \in M^*$, we form $K = \{(f_1(m),\ldots,f_n(m)) \mid m \in M\}$ which is a submodule of ${}_RE^n$. We claim that $(h(f_1),\ldots,h(f_n)) \in K$. Otherwise by Proposition 1.12 again, there exists some $s = \oplus s_i : E^n \longrightarrow E$ such that $s(K) = 0$ but $s(h(f_1),\ldots,h(f_n)) \neq 0$, where each $s_i \in S$. Hence $0 = s(f_1(m),\ldots,f_n(m)) = (\Sigma s_i f_i)(m)$, for each $m \in M$. So $\Sigma s_i f_i = 0$. Since h is an S-homomorphism, we have $0 \neq s(h(f_1),\ldots,h(f_n)) = \Sigma s_i(h(f_i)) = h(\Sigma s_i f_i) = h(0) = 0$, a contradiction. This establishes our claim. Hence $h(f_i) = f_i(m_{f_1,\ldots,f_n})$ for some $m_{f_1,\ldots,f_n} \in M$. In other words, M has a finitely solvable family $\{m_f, \mathrm{Ker}(f)\}_{f \in M^*}$. By the linear compactness of M, there is some $m \in M$ such that $f(m-m_f) = 0$ for all $f \in M^*$. So $f(m) = f(m_f) = h(f)$ for all $f \in M^*$. Hence $h = \sigma_M(m)$, and σ_M is epic.

If $L \leqq M^*_S$ and $h \in \mathrm{Hom}_S(L, E)$, one can see similarly as above that there exists an element $m \in M$ such that $h(f) = f(m)$ for all $f \in L$. Define $g \in M^{**}$ via $g(f) = f(m)$, then $g|_L = h$.

(2)==>(1). If $\{m_i, M_i\}_{i \in I}$ is a finitely solvable family of M, then the map

$$h\colon \sum_{i \in I} \mathrm{Ran}_{M^*}(M_i) \longrightarrow E_S$$

via

$$\Sigma f_i \longmapsto \Sigma f_i(m_i)$$

is well-defined. By assumption we have an S-homomorphism $g \in M^{**}$ which extends h. Now M is E-reflexive, so $g = \sigma_M(m)$ for some $m \in M$. Hence $m - m_i \in \mathrm{Lan}_M \mathrm{Ran}_{M^*}(M_i) = M_i$ for each i, since ${}_RE$ is a cogenerator. (If $\mathrm{Lan}_M \mathrm{Ran}_{M^*}(M_i) \supset M_i$ for some i, then we apply Proposition 1.12 to derive a contradiction.) Consequently M is linearly compact.

COROLLARY 4.2 (Müller). If ${}_RE_S$ defines a duality, then the E-reflexive left R-modules (right S-modules) are precisely the linearly compact left R-modules (right S-modules).

PROOF. Since ${}_RE_S$ is faithfully balanced (Theorem 2.6), and both ${}_RE$ and E_S are injective cogenerators, the results follow from Theorem 4.1.

If ${}_RE_S$ defines a duality, then $R \cong \mathrm{End}(E_S)$ and $S \cong \mathrm{End}({}_RE)$ canonically. For convenience, we say that R has a duality (induced by ${}_RE$) or that ${}_RE$ defines a duality. The following Müller's theorem is a criterion for an arbitrary ring to have a duality.

THEOREM 4.3 (Müller). A ring R has a duality if and only if both ${}_RR$ and the minimal injective cogenerator ${}_RU$ in R-Mod are linearly compact, and then ${}_RU$ induces such a duality.

PROOF. (==>). Let ${}_RE$ define a duality. Since ${}_RR$ and ${}_RE$ are E-reflexive, they are linearly compact by Theorem 4.1. Hence ${}_RU$ is linearly compact since it can be embedded into ${}_RE$.

(<==). Since ${}_RU$ is linearly compact, by Proposition 3.4, ${}_RU = \bigoplus_{i=1}^{n} E(T_i)$ which is injective, where $T_1,\ldots,T_n$ is an irredundant set of simple left R-modules. Hence $S = \mathrm{End}({}_RU)$ is a semiperfect ring by Proposition 1.10. By Theorem 4.1 (using the linear compactness of ${}_RU$ again), U_S is S-injective, so is injective. Let $e_i : U \longrightarrow E(T_i)$ be the projection. Then $e_1,\ldots,e_n$ is a basic set in S and any simple right S-module is of the form e_iS/e_iJ for some e_i, where $J = J(S)$. Now $T_iJ = 0$ (Proposition 1.11), and $T_ie_i = T_i = Rt_i \neq 0$. Let

$$h\colon e_iS/e_iJ \longrightarrow T_i = Rt_i \leq U$$

via

$$e_is+e_iJ \longmapsto t_ie_is.$$

Then h is well-defined since $T_iJ = 0$, and then it is a non-zero right S-homomorphism. Hence h is monic since e_iS/e_iJ is simple. Consequently, U_S contains all simple right S-modules, so U_S is a cogenerator by Proposition 1.13. Clearly S_S is U-reflexive and ${}_RR$ is U-reflexive by Theorem 4.1. Hence ${}_RU_S$ is faithfully balanced by Theorem 1.20. It follows from Theorem 2.4 that R has a duality induced by ${}_RU_S$.

It follows from Theorem 4.3 and Corollary 3.14 that a ring with a duality must be semiperfect.

The next result follows from Corollary 3.5 and the theorem.

COROLLARY 4.4 (Azumaya, Morita). A left artinian ring R has a duality if and only if every indecomposable injective left R-module (i.e., the injective envelope of a simple left R-module) is finitely generated.

The next result was mentioned in Vámos [77] as a slightly modified version of Müller's Theorem 4.3.

THEOREM 4.5. Let R be a ring and let ${}_RE \in$ R-Mod. Then the following statements are equivalent:

(1) ${}_RE$ defines a duality;

(2) ${}_RR$ is linearly compact and ${}_RE$ is a linearly compact and finitely cogenerated injective cogenerator.

PROOF. Let $S = \mathrm{End}({}_RE)$.

(==>). By Theorem 4.3, ${}_RR$ is linearly compact. Since ${}_RE$ is a cogenerator which is E-reflexive and E_S is injective, ${}_RE$ is linearly compact by Theorem 4.1.

(<==). This is similar to that given in Theorem 4.3.

A ring R has (<u>Morita</u>) <u>self-duality</u> in case there exists an R-bimodule ${}_RE_R$ that defines a Morita duality. Hence a ring R has self-duality if and only if there is an R-module ${}_RE$ that defines a Morita duality such that $R \cong \mathrm{End}({}_RE)$.

An immediate consequence of Theorem 4.5 is that Morita

duality and self-duality are preserved by equivalence. Recall that a ring with a duality must be semiperfect and any semiperfect ring is equivalent to its basic ring. Hence to show that a ring R has a duality, we may assume that R is a basic semiperfect ring according to the following corollary.

COROLLARY 4.6. Let R and S be equivalent rings via the bimodule ${}_RP_S$. Then

(1) If ${}_RE_T$ defines a duality, then ${}_S\mathrm{Hom}_R(P,E)_T$ defines a duality;

(2) If ${}_RE_R$ defines a self-duality, then ${}_S\mathrm{Hom}_R(Q, \mathrm{Hom}_R(P,E))_S$ defines a self-duality, where $Q = \mathrm{Hom}_R(P,R)$;

(3) If ${}_RE_T$ and ${}_RE'_S$ define dualities, then ${}_T\mathrm{Hom}_R(E,E')_S$ defines an equivalence.

PROOF. Let ${}_S(P^*)_T = {}_S\mathrm{Hom}_R(P,E)_T$. Since ${}_RE$ is a finitely cogenerated injective cogenerator, so is ${}_SP^*$ by Proposition 1.14. Since ${}_RE$ and ${}_RR$ are linearly compact, so are ${}_SP^*$ and ${}_SS = \mathrm{Hom}_R(P,P)$ by Corollaries 3.17 and 3.18. Hence ${}_SP^*$ defines a duality between left S-modules and right $\mathrm{End}({}_SP^*)$ ($\cong \mathrm{End}({}_RE) = T$) -modules, and so (1) follows.

By (1), ${}_S(P^*)_R$ defines a duality. Now ${}_SQ_R$ also defines an equivalence, so by the right hand version of (1) we conclude that ${}_S\mathrm{Hom}_R(Q,P^*)_S$ defines a duality. This completes the proof of (2).

For (3), since ${}_RE$ is a finitely cogenerated injective cogenerator, $\mathrm{Hom}_R(E,E')_S$ is a finitely generated projective generator (Here each simple right S- and left R-module is E'-reflexive and each simple left R-module can be embedded into ${}_RE$, it follows that each simple right S-module is an epimorphic image of $\mathrm{Hom}_R(E,E')_S$, and then $\mathrm{Hom}_R(E,E')_S$ is a generator.) Now $\mathrm{End}(\mathrm{Hom}_R(E,E')_S) \cong \mathrm{End}({}_RE) \cong T$, hence ${}_T\mathrm{Hom}_R(E,E')_S$ induces an equivalence.

COROLLARY 4.7. Let $R = R_1 \oplus \ldots \oplus R_n$. Then (1) R has a duality if and only if each R_i has a duality; and (2) R has self-duality if each R_i has self-duality.

PROOF. (1) (==>) This follows from Corollary 2.5. For (1) (<==) and (2): If ${}_{R_i}(E_i)_{S_i}$ (${}_{R_i}(E_i)_{R_i}$) defines a duality, let $E = \oplus_{i=1}^{n} E_i$ and $S = \oplus_{i=1}^{n} S_i$. Then it is immediate that ${}_R E_S$ (${}_R E_R$) is a bimodule that defines a duality.

Using the profound results in Dowbor, Ringel and Simson [80] and Schofield [85b, p.214-217], Kraemer [87] has constructed an artinian ring $R = R_1 \oplus \ldots \oplus R_5$ with self-duality but none of R_i's has self-duality. Hence, the converse of Corollary 4.7(2) is not true. This also answers the question of Haack [79, p.359] in the negative.

The following important result is due to Müller [70].

THEOREM 4.8. If R is a commutative ring with a duality, then R has self-duality.

PROOF. Since R is a commutative semiperfect ring (Theorem 2.7), it is a finite direct sum of local rings (Theorem 1.18). We may assume that R itself is a local ring (Corollary 4.7), and suppose that the minimal cogenerator ${}_R U$ induces a duality of R with $S = \mathrm{End}({}_R U)$. Since R is local, ${}_R U = E({}_R T)$ where ${}_R T = R/J$ $(J = J(R))$ is the unique simple left R-module. We view R as a subring of S, since R is commutative.

For any $s \in S$, there exists a maximal S-submodule V of U such that $V(s-r) = 0$ for a suitable $r \in R$. In fact, if $\{V_k\}_k$ is a chain of S-submodules such that $V_k(s-r_k) = 0$, then R has a finitely solvable family $\{r_k, \mathrm{Lan}_R(V_k)\}_k$ that is solvable since R is linearly compact. Let $r \in R$ be a solution, then $(\cup_k V_k)(s-r) = 0$. Hence $(\cup_k V_k)$ is an upper bound for the chain $\{V_k\}_k$. Now Zorn's Lemma provides the maximal V. Note $V \neq 0$, since the simple R-submodule T of U is also an S-submodule of U and satisfies $T(s-r) = 0$ for some $r \in R$. Hence $\mathrm{Lan}_R(V) = I \neq R$ by Theorem 2.6. So $I \leq J$ since R is local. Since R is commutative, by Theorem 2.6 again we have

$$\begin{aligned}\mathrm{Ran}_S(V) &= \mathrm{Ran}_S(\mathrm{Ran}_U(\mathrm{Lan}_R(V))) = \mathrm{Ran}_S(\mathrm{Ran}_U(I)) \\ &= \mathrm{Ran}_S(\mathrm{Lan}_U(I)) = \mathrm{Ran}_S(\mathrm{Lan}_U(IS)) = IS.\end{aligned}$$

Hence $s-r = \Sigma_k\, a_k s_k$ for some $a_k \in I \leq J$ and $s_k \in S$. So $S = R+JS$, then $s_k = r_k + \Sigma j_{il_i} s_{il_i}$ for some $r_k \in R$, $j_{il_i} \in J$, $s_{il_i} \in S$; and

$$s-r-\Sigma r_k a_k = \Sigma a_k s_k - \Sigma r_k a_k = \Sigma a_k(s_k - r_k) = \Sigma a_k(\Sigma j_{il_i} s_{il_i}) \in (\Sigma a_k J)S.$$

If $\Sigma a_k R \neq 0$, then $\Sigma a_k J \subset \Sigma a_k R \leq I$. Setting $\mathrm{Ran}_U(\Sigma a_k J) = W \supset \mathrm{Ran}_U(I) = V$ (Theorem 2.6), we get $W(s-r-\Sigma r_k a_k) = 0$ contradicting to the maximality of V. Hence $\Sigma a_k R = 0$. Then $a_k = 0$, $s = r$ and $S = R$.

The next consideration reduces the study of dualities, to a certain extent, from semiperfect rings to local rings.

LEMMA 4.9. Let R be a ring with $1 = e_1 + \ldots + e_n$ where e_i's are idempotents. Then an R-module $_RM$ is linearly compact if and only if each $_{e_iRe_i}e_iM$ is linearly compact.

PROOF. (==>). Let $_RM$ be linearly compact and e be an idempotent of R. We note that each eRe-submodule of eM is of the form eN, where N is an R-submodule of M. We assume that $\{em_j, eM_j\}_j$ is a finitely solvable family of $_{eRe}eM$, where each $M_j \leq {}_RM$. Then $\{em_j, M_j\}_j$ is a finitely solvable family of the linearly compact module $_RM$. If $m \in M$ is a solution for $\{m_j, M_j\}_j$ then em is a solution for $\{em_j, eM_j\}_j$. Hence $_{eRe}eM$ is linearly compact.

(<==). Let $\{m_j, M_j\}_j$ be a finitely solvable family of $_RM$.

Then $\{e_i m_j, e_i M_j\}_j$ is a finitely solvable family of $e_i M$ $(i=1,\ldots,n)$. Now there exists $x_i \in e_i M$ such that $x_i - e_i m_j \in e_i M$ for all j $(i=1,\ldots,n)$. If we take $x = x_1 + \ldots + x_n \in M$ then $x - m_j = \Sigma_{i=1}^n (x_i - e_i m_j) \in M$ for all j. This shows that ${}_R M$ is linearly compact.

Fuller [69] proved the following useful result which we also need in Chapters 3 and 4.

LEMMA 4.10. Let ${}_R E$ be an injective R-module and e be an idempotent of R such that $\mathrm{Ran}_E(eR) = 0$. Then, for each left R-module M, restriction to eM gives an abelian group isomorphism

$$\mathrm{Hom}_R(M,E) \cong \mathrm{Hom}_{eRe}(eM,eE).$$

Moreover,

(1) The map f defined via

$$[f(s)](ex) = s(ex),\ s \in \mathrm{Hom}_R(E,E),\ ex \in eE,$$

is a ring isomorphism from $\mathrm{Hom}_R(E,E)$ to $\mathrm{Hom}_{eRe}(eE,eE)$;

(2) If $I \subseteq K$ are ideals of R then

$$\mathrm{Ran}_E(I)/\mathrm{Ran}_E(K) \cong \mathrm{Hom}_{eRe}(eK/eI,eE)$$

as left R-modules.

(3) The eRe-module ${}_{eRe}eE$ is injective.

PROOF. If $m^* \in \mathrm{Hom}_R(M,E)$ then, for all $r \in R$, $m \in M$,

$$m^*(erem) = ere \cdot m^*(em) \in eE.$$

Thus $m^*|_{eM} \in \mathrm{Hom}_{eRe}(eM,eE)$ and the restriction to eM is an additive map from $\mathrm{Hom}_R(M,E)$ to $\mathrm{Hom}_{eRe}(eM,eE)$. If $0 = m^*(eM) = eRm^*(M)$, then m^* is the zero map since $\mathrm{Ran}_E(eR) = 0$. Let $g \in \mathrm{Hom}_{eRe}(eM,eE)$. If $r_i \in R$, $m_i \in M$ with $\Sigma\, r_i em_i = 0$, then for each $er \in eR$,

$$er(\Sigma\, r_i g(em_i)) = \Sigma\, err_i g(em_i) = g(er(\Sigma\, r_i em_i) = g(0) = 0.$$

So $\Sigma\, r_i g(em_i) \in \mathrm{Ran}_E(eR) = 0$. Thus there exists an R-homomorphism $\bar{g} : ReM \longrightarrow E$ defined via

$$\bar{g}(\Sigma\, r_i em_i) = \Sigma\, r_i g(em_i).$$

Using the injectivity of E, let $\hat{g} : M \longrightarrow E$ extend $\bar{g}$. Then $\hat{g}(em) = g(em)$ for all $em \in eM$ and we have shown that restriction to eM provides the desired isomorphism $\mathrm{Hom}_R(M,E) \cong \mathrm{Hom}_{eRe}(eM,eE)$.

To see (1) holds take $M = E$.

For (2) note that, since E is injective,

$$[h(x + \mathrm{Ran}_E(K))](k + I) = kx,\ x \in \mathrm{Ran}_E(I),\ k \in K$$

defines an isomorphism

$$h : \mathrm{Ran}_E(I)/\mathrm{Ran}_E(K) \longrightarrow \mathrm{Hom}_R(K/I,E).$$

But we know that

$$\mathrm{Hom}_R(K/I,E) \cong \mathrm{Hom}_{eRe}(eK/eI,eE).$$

Now it is routine to check that these isomorphisms commute with left multiplication by elements of R.

(3) Clearly, every left ideal of the ring eRe is of the form eIe where Ie is a left ideal of R. If $g : eIe \longrightarrow eE$ is an eRe-homomorphism then g can be extended to an R-homomorphism $\bar{g} : Ie \longrightarrow E$. Since E is injective there is an $x \in E$ with $\bar{g}(a) = ax$ for all $a \in Ie$. Thus $ex \in eE$ and $g(eae) = eae\cdot ex$ for all $eae \in eIe$. Hence ${}_{eRe}eE$ is an injective eRe-module.

LEMMA 4.11. Let R be a semiperfect ring with radical J. If e is an idempotent of R then $eE({}_R Re/Je) \cong E({}_{eRe}eRe/eJe)$ as left eRe-modules.

PROOF. Let $E = E({}_R Re/Je)$, then $\mathrm{Ran}_E(eR) = 0$. Since ${}_R E$ has cyclic essential socle Re/Je, ${}_{eRe}eE$ has cyclic essential socle $e(Re/Je) \cong eRe/eJe$. Then

$$E({}_{eRe}eRe/eJe) \cong E({}_{eRe}eE) = eE$$

by Lemma 4.10(3), where the first isomorphism follows from the fact that if a module M has an essential submodule N then $E(N) \cong E(M)$.

Now let R be a semiperfect ring with a complete set of (primitive) idempotents $e_1, \ldots, e_n$. Then R is naturally isomorphic to the matrix ring $(e_iRe_j)_{ij}$ via $r \longmapsto (e_ire_j)_{ij}$. In the following, Müller [84] characterized the duality of R via the dualities of e_iRe_i's (the local rings e_iRe_i's).

THEOREM 4.12. The semiperfect ring R has a duality if and only if each e_iRe_i has a duality (with S_i induced by bimodule E_i) and each e_iRe_j and $e_iRe_j^* = \mathrm{Hom}_{e_iRe_i}(e_iRe_j, E_i)$ is linearly compact as a left module. Moreover, the corresponding second ring is $S = (e_iRe_j^{**})_{ij}$.

PROOF. Let $J = J(R)$, then ${}_RE = E({}_RR/J) = \oplus_{i=1}^{n} E(Re_i/Je_i)$ is a finitely cogenerated injective cogenerator.

(==>). By Theorem 4.5, ${}_RR$ and ${}_RE$ are linearly compact. Since left R-modules Re_j and $E(Re_i/Je_i)$ are linearly compact, the left e_iRe_i-modules e_iRe_j and $e_iE(Re_i/Je_i)$ are linearly compact by Lemma 4.9. As left e_iRe_i-modules, Lemma 4.11 gives

$$e_iE(Re_i/Je_i) \cong E({}_{e_iRe_i}e_iRe_i/e_iJe_i)$$

which is a finitely cogenerated injective cogenerator of the (local) ring e_iRe_i. Hence e_iRe_i has a duality by Theorem 4.5 again. Finally as left e_jRe_j-modules, we have

$$\begin{aligned}(e_iRe_j)^* &= \mathrm{Hom}_{e_iRe_i}(e_iRe_j, e_iE(Re_i/Je_i)) \\ &\cong \mathrm{Hom}_R(Re_j, E(Re_i/Je_i)) \\ &\cong e_jE(Re_i/Je_i),\end{aligned}$$

where the first isomorphism follows from Lemma 4.10 since we have

$\mathrm{Ran}_{E(Re_i/Je_i)}(e_iR) = 0$, and the last isomorphism is given by $f \longmapsto f(e_j)$. Hence $(e_iRe_j)^* \cong e_jE(Re_i/Je_i)$ is linearly compact by Lemma 4.9.

(<==). Since each left e_iRe_i-module e_iRe_j is linearly compact, the left R-module Re_j is linearly compact by Lemma 4.9, hence ${}_RR = \oplus_{j=1}^{n} Re_j$ is linearly compact. Now we show that ${}_RE$ is linearly compact. Since ${}_{e_iRe_i}E_i$ defines a duality, it is a finitely cogenerated injective cogenerator. Now ${}_{e_iRe_i}e_iE(Re_i/Je_i)$ is also a finitely cogenerated injective cogenerator, so it is linearly compact and defines a duality. Hence we may assume that $E_i = e_iE(Re_i/Je_i)$ as e_iRe_i-modules. By the above proof, we see that, as left e_jRe_j-modules, $e_jE(Re_i/Je_i) \cong (e_iRe_j)^*$, which is linearly compact by hypothesis. It follows that each ${}_RE(Re_i/Je_i)$ is linearly compact by Lemma 4.9, and then ${}_RE$ is linearly compact.

If R has a duality induced by ${}_RE = \oplus_{i=1}^{n} E(Re_i/Je_i)$ then the corresponding second ring is $S = \mathrm{End}({}_RE)$, which is isomorphic to the matrix ring $(\mathrm{Hom}_R(E_i,E_j))_{i,j}$, where $E_i = E(Re_i/Je_i)$. Now

$$e_iRe_j^{**} = (e_iRe_j^*)^* \cong (e_jE_i)^*$$

$$= \mathrm{Hom}_{e_jRe_j}(e_jE_i, e_jE_j) \cong \mathrm{Hom}_R(E_i,E_j).$$

Hence we can view the second ring as $S = (e_iRe_j^{**})_{ij}$, where the multiplication between $e_iRe_j^{**}$ and $e_jRe_k^{**}$ is what is induced by the composition of $\mathrm{Hom}_R(E_i,E_j)$ and $\mathrm{Hom}_R(E_j,E_k)$.

The following result is due to Miller and Turnidge [72].

COROLLARY 4.13. If R has a duality and ${}_RP$ is a finitely generated projective R-module, then the endomorphism ring $\mathrm{End}({}_RP)$ has a duality.

PROOF. Suppose that the semiperfect ring R has a complete set of primitive idempotents $e_1,\ldots,e_n$. Since ${}_RP$ is finitely generated and projective, ${}_RP \cong (Re_{i_1})^{k_1}\oplus\ldots\oplus(Re_{i_m})^{k_m}$, where $k_1,\ldots,k_m$ are positive integers and $1 \leq i_1 <\ldots< i_k \leq n$. (For a proof, see Anderson-Fuller [74].) Then $\mathrm{End}({}_RP)$ is equivalent to its basic ring eRe where $e = e_{i_1}+\ldots+e_{i_m}$. The ring eRe has a duality by Theorem 4.12, and it follows from Corollary 4.6 that $\mathrm{End}({}_RP)$ also has a duality.

COROLLARY 4.14. If R has a duality and ${}_RU$ is a finitely cogenerated injective R-module, then the endomorphism ring $\mathrm{End}({}_RU)$ has a right duality.

PROOF. Let ${}_RE_S$ define a duality. Then $U^*_S = \mathrm{Hom}_R(U,E)_S$ is a finitely generated projective S-module by Theorem 2.6. It follows from the right version of Corollary 4.13 that $\mathrm{End}(U^*_S)$ has a right duality, so does $\mathrm{End}({}_RU)$.

Generalizations of the above corollaries have been given in a recent paper by J.L. Gómez Pardo and N. Rodriguez Gonzalez [91].

Now we conclude this section by asking the following two questions:

QUESTION 4.15. Let ${}_RE_S$ define a duality. Then ${}_RR$ and S_S are linearly compact. If R_R is linearly compact, is ${}_SS$ linearly compact?

QUESTION 4.16. If R has a duality, does the formal power series ring $R[[x]]$ have a duality?

SECTION 5. COGENERATOR RINGS

A ring R is called a <u>cogenerator ring</u> in case ${}_RR_R$ defines a duality. Clearly any semisimple ring is a cogenerator ring. We shall give non-trivial examples of cogenerator rings in Section 10. In this section, several characterizations of cogenerator rings are presented.

An R-module ${}_RM$ is called <u>F-injective</u> if every homomorphism from a finitely generated left ideal I of R to M is given by right multiplication by an element of M, i.e., the homomorphism can be extended to R. If R is a left noetherian ring then F-injective left R-modules are injective. It is easy to see that an arbitrary direct sum of F-injective modules is F-injective. Hence, if R is not left noetherian then there is a left F-injective R-module that is not injective, since a ring R is left noetherian if and only if every direct sum of injective left R-modules is injective (see Anderson-Fuller [74, Proposition 18.13]).

PROPOSITION 5.1 (Ikeda and Nakayama [53]) Let R be a ring. Then ${}_RR$ is F-injective if and only if R satisfies

(1) $\mathrm{Ran}_R(\mathrm{Lan}_R(a)) = aR$ for each $a \in R$; and

(2) $\mathrm{Ran}_R(I_1 \cap I_2) = \mathrm{Ran}_R(I_1) + \mathrm{Ran}_R(I_2)$ for all finitely generated left ideals I_1, I_2.

PROOF. (==>). Let $a \in R$. We have $aR \subseteq \mathrm{Ran}_R(\mathrm{Lan}_R(a))$. If $b \in \mathrm{Ran}_R(\mathrm{Lan}_R(a))$, one checks that $f : Ra \longrightarrow Rb$ via $ra \longmapsto rb$ is an R-homomorphism. Since ${}_RR$ is F-injective, there is some $c \in R$ such that $f(ra) = rac$ for all $ra \in Ra$. In particular, $b = f(a) = ac \in aR$, hence $\mathrm{Ran}_R(\mathrm{Lan}_R(r)) \subseteq aR$ and then (1) holds. For (2), it is trivial we have $\mathrm{Ran}_R(I_1 \cap I_2) \supseteq \mathrm{Ran}_R(I_1) + \mathrm{Ran}_R(I_2)$. So let $r \in \mathrm{Ran}_R(I_1 \cap I_2)$. Let $f : I_1+I_2 \longrightarrow R$ via $a_1+a_2 \longmapsto a_1r$. If $a_1+a_2 = 0$, $a_1 \in I_1 \cap I_2$ and so $a_1r = 0$. Hence f is well-defined and then f is an R-homomorphism. By hypothesis, there exists an element $c \in R$ with $a_1r = f(a_1+a_2) = (a_1+a_2)c$ for all $a_1+a_2 \in I_1+I_2$. In particular, $0 = f(a_2) = a_2c$ for all $a_2 \in I_2$. Hence $c \in \mathrm{Ran}_R(I_2)$. For all $a_1 \in I_1$ we have $a_1r = f(a_1) = a_1c$. So

$r-c \in \mathrm{Ran}_R(I_1)$. It follows that $r = (r-c)+c \in \mathrm{Ran}_R(I_1) + \mathrm{Ran}_R(I_2)$. Therefore $\mathrm{Ran}_R(I_1 \cap I_2) \subseteq \mathrm{Ran}_R(I_1) + \mathrm{Ran}_R(I_2)$, and then the equality in (2) holds.

(<==). To show that ${}_RR$ is F-injective, we consider an R-homomorphism $f : I \longrightarrow R$ where I is a finitely generated left ideal, say $I = Ra_1+\ldots+Ra_n$. We use induction on n to prove that there is some $c \in R$ with $f(r) = rc$ for all $r \in I$.

Let $I = Ra$. If $ra = 0$, then $0 = f(ra) = rf(a)$. Hence $\mathrm{Lan}_R(a) \subseteq \mathrm{Lan}_R(f(a))$. So

$$aR = \mathrm{Ran}_R(\mathrm{Lan}_R(a)) \supseteq \mathrm{Ran}_R(\mathrm{Lan}_R(f(a)) = f(a)R.$$

Then $f(a) = ac$ for some $c \in R$, and it follows that $f(ra) = rf(a) = rac$ for all $ra \in Ra$. So it is true for $n = 1$.

Now let $I = Ra_1+\ldots+Ra_{n-1}+Ra_n$ and suppose that we have $c_1, c_2 \in R$ with

$$f(\Sigma_{i=1}^{n-1} r_i a_i) = (\Sigma_{i=1}^{n-1} r_i a_i)c_1 \text{ and } f(r_n a_n) = r_n a_n c_2.$$

If $b \in (Ra_1+\ldots+Ra_{n-1}) \cap Ra_n$, then $bc_1 = f(b) = bc_2$ and so $c_1-c_2 \in \mathrm{Ran}_R((Ra_1+\ldots+Ra_{n-1}) \cap Ra_n) = \mathrm{Ran}_R(Ra_1+\ldots+Ra_{n-1}) + \mathrm{Ran}_R(Ra_n)$. Let $c_1-c_2 = d_1+d_2$, where $d_1 \in \mathrm{Ran}_R(Ra_1+\ldots+Ra_{n-1})$ and $d_2 \in \mathrm{Ran}_R(Ra_n)$. Let $c = c_1-d_1 = c_2+d_2$. Then we have

$$\begin{aligned} f(\Sigma_{i=1}^{n} r_i a_i) &= f(\Sigma_{i=1}^{n-1} r_i a_i) + f(r_n a_n) \\ &= (\Sigma_{i=1}^{n-1} r_i a_i)(c_1-d_1) + (r_n a_n)(c_2+d_2) \\ &= (\Sigma_{i=1}^{n-1} r_i a_i)c + r_n a_n c = (\Sigma_{i=1}^{n} r_i a_i)c. \end{aligned}$$

Although an F-injective module need not be injective, we do have the following

PROPOSITION 5.2. Let R be a right linearly compact ring. If ${}_RR$ is F-injective, then ${}_RR$ is injective.

PROOF. Let $f : I \longrightarrow {}_RR$ be a left R-homomorphism where I is a left ideal of R. Since ${}_RR$ is F-injective, for each finitely generated left ideal $I_i \leq I$ there exists an element $r_i \in R$ such that $f(a_i) = a_i r_i$ for all $a_i \in I_i$. It is easy to see that the right R-module R_R has a finitely solvable family $\{m_i, \mathrm{Ran}_M(I_i)\}_i$. Since R_R is linearly compact, we obtain an element $r \in R$ such that $r - r_i \in \mathrm{Ran}_R(I_i)$ for all i. For any $a \in I$, let $I_i = Ra \leq I$, we have $a(r - r_i) = 0$ and then $f(a) = ar_i = ar$. Therefore by Baer's lemma $_RR$ is injective.

A ring R is said to have the <u>double annihilator property</u> if

$$\mathrm{Lan}_R\mathrm{Ran}_R(I) = I \quad \text{and} \quad \mathrm{Ran}_R\mathrm{Lan}_R(L)$$

for each left ideal I and each right ideal L.

PROPOSITION 5.3. If R has the double annihilator property, then both $_RR$ and R_R are F-injective.

PROOF. Let I_1 and I_2 be any two left ideals of R. We have

$$\begin{aligned}\mathrm{Lan}_R\mathrm{Ran}_R(I_1 \cap I_2) &= I_1 \cap I_2 \\ &= \mathrm{Lan}_R\mathrm{Ran}_R(I_1) \cap \mathrm{Lan}_R\mathrm{Ran}_R(I_2) \\ &= \mathrm{Lan}_R(\mathrm{Ran}_R(I_1) + \mathrm{Ran}_R(I_2))\end{aligned}$$

Then we obtain $\mathrm{Ran}_R(I_1 \cap I_2) = \mathrm{Ran}_R(I_1) + \mathrm{Ran}_R(I_2)$. Hence by Proposition 5.1, $_RR$ is F-injective. By symmetry, R_R is F-injective.

Now we give several characterizations of cogenerator rings. One notes that the implication "(3) ==> (2)" follows from a result of Colby and Fuller [84] who have shown that if $_RE_S$ is any balanced bimodule such that $_RE$ and E_S are cogenerators then $_RE$ and E_S are injective.

THEOREM 5.4. The following are equivalent for a ring R:

(1) R is a cogenerator ring;

(2) $_RR$ and R_R are injective cogenerotors;

(3) ${}_RR$ and R_R are cogenerators;

(4) ${}_RR$ and R_R are injective and R has the double annihilator property;

(5) ${}_RR$ and R_R are linearly compact and R has the double annihilator property;

(6) All finitely generated left and right R-modules are R-reflexive;

(7) All cyclic left and right R-modules are R-reflexive.

PROOF. (1) <==> (2). Theorem 2.4.

(3) ==> (2). By (3) there is a set A with ${}_RR \leqq E = E({}_RR) \leqq {}_RR^A$. Let $p_a: R^A \longrightarrow R$ be the a-th projection for each $a \in A$. Let $p_a(1) = e_a$ and $I = \Sigma_A e_aR \leqq R_R$. Then $Lan_R(I) = 0$. Since R_R is a cogenerator, by Proposition 1.12 we get $I = R$. Say $1 = \Sigma_A e_ar_a$ with almost all r_a zero. Then $x \longmapsto \Sigma_A p_a(x)r_a$ is a spilit epimorphism $E \longrightarrow {}_RR$. It follows that ${}_RR$ is a summand of E and so ${}_RR$ is injective. Similarly, R_R is injective.

(1), (2) ==> (4). Since ${}_RR_R$ induces a duality, R has the double annihilator property by Theorem 2.6(2).

(4) ==> (3). Let R/I be a simple left R-module, i.e., I is a maximal left ideal. Since $Lan_RRan_R(I) = I \neq R$, $Ran_R(I) \neq 0$. Let $0 \neq a \in R$ with $Ia = 0$. Then $f : R/I \longrightarrow R$ via $r + I \longmapsto ra$ is well-defined hence an R-homomorphism. Now f is not an zero map and R/I is simple, f is an monomorphism. Hence ${}_RR$ is a cogenerotor by Proposition 1.13. Similarly, R_R is a cogenerotor.

(1), (4) ==> (5). This follows from Theorem 4.5.

(5) ==> (4). Propositions 5.2 and 5.3.

(1) ==> (6). Theorem 2.6(9).

(6) ==> (7). Obvious.

(7) ==> (1). Theorem 2.4(2).

Recently, Ánh [91] gives the following characterization of cogenerator rings, using the methods developed in his earlier paper [90] which we shall presented in the next section.

THEOREM 5.5 (Ánh [91]). A ring R is a cogenerator ring if and only if the modules ${}_RR$ and R_R are both linearly compact and finitely cogenerated and the ${}_RR_R$-dual takes simples to simples.

A ring R is called a <u>left</u> (<u>right</u>) <u>PF-ring</u> (pseudo-Frobenius) in case ${}_RR$ (R_R) is an injective cogenerator. Hence R is a cogenerator ring if and only if R is both left and right PF-ring. Dischinger and Müller [86] has constructed a left PF-ring that is not right PF (see also Kraemer [87b]). For references to the next proposition, consult Azumaya [66], Osofsky [66], and Utumi [67]. The proof is also given in Faith [76, Proposition 24.32].

PROPOSITION 5.6. The following statements are equivalent for a ring R:

(1) R is a left PF-ring;

(2) Every faithful left R-module is a generator;

(3) ${}_RR$ is an injective module with finitely generated essential socle;

(4) R is a semiperfect ring and ${}_RR$ is an injective module with essential socle.

SECTION 6. DUALITY OF LINEARLY COMPACT COMMUTATIVE RINGS

It was conjectured that every linearly compact commutative ring has a Morita duality (see Zelinsky [53] and Muller [70]). This has been verified for artinian rings (Azumaya [59], Morita [58]), and for noetherian rings (Matlis [58]). Recently, Ánh [90] has settled this conjecture in the affirmative. We introduce

Anh's result [90] in this section. So R will always be a commutative ring throughout this section, $M^* = \mathrm{Hom}_R(M,R)$ for an R-module M, and write $\mathrm{Ann}(I) = \mathrm{Lan}_R(I)$ for and subset $I \subseteq R$.

R is said to be <u>subdirectly irreducible</u> in case the intersection H of all non-zero ideals of R is not zero. Then H is obviously the smallest non-zero ideal of R. If R is a subdirectly irreducible linearly compact ring then since R is semiperfect, R is a local ring with essential minimal ideal H; in particular R is finitely cogenerated. An ideal I of R is called <u>sheltered</u> if the factor ring R/I is subdirectly irreducible.

We need a series of preparations to establish the main result.

PROPOSITION 6.1. Let $\{I_i\}$ be an inverse system of ideals in a linearly compact subdirectly irreducible ring R. Then the equality $\mathrm{Ann}(\cap I_i) = \cup \mathrm{Ann}(I_i)$ holds.

PROOF. The inclusion $\mathrm{Ann}(\cap I_i) \supseteq \cup \mathrm{Ann}(I_i)$ is obvious. Since R is commutative, the multiplication with $r \in R$ is an R-homomorphism of ${}_RR$ hence $r(\cap I_i) = \cap rI_i$ by Theorem 3.8. Now let $r \in \mathrm{Ann}(\cap I_i)$, then $0 = r(\cap I_i) = \cap rI_i$. Since R is subdirectly irreducible, $rI_j = 0$ for some j and so $r \in \cup \mathrm{Ann}(I_i)$. Therefore $\mathrm{Ann}(\cap I_i) \subseteq \cup \mathrm{Ann}(I_i)$. This shows the validity of the equality $\mathrm{Ann}(\cap I_i) = \cup \mathrm{Ann}(I_i)$.

PROPOSITION 6.2. For any non-zero finitely generated ideal I of a linearly compact subdirectly irreducible ring R with maximal ideal J, we have $\mathrm{Ann}(I) < \mathrm{Ann}(JI)$.

PROOF. For $I = Ra$ there is an element $b \in R$ such that ba generates the smalleat non-zero ideal H of R and hence $b \in \mathrm{Ann}(Ja)\setminus\mathrm{Ann}(Ra)$, i.e., $\mathrm{Ann}(Ja) > \mathrm{Ann}(Ra)$. Assume now by induction that $I = Ra_1+\ldots+Ra_n$ and for $I_1 = Ra_1+\ldots+Ra_{n-1}$ we have $\mathrm{Ann}(I_1) < \mathrm{Ann}(JI_1)$. Consider an element

$b \in Ann(JI_1)\setminus Ann(I_1)$. If $ba_n = 0$, then we are clearly done. For the case $ba_n \neq 0$ there is an element $r \in R$ such that rba_n generates H. This shows that $rb \in Ann(JI)\setminus Ann(I)$, and hence our proof is complete.

PROPOSITION 6.3. For any non-zero finitely generated module M over a linearly compact subdirectly irreducible ring R, its dual M^* is finitely cogenerated.

PROOF. For any $0 \neq f \in M^* = Hom_R(M,R)$ the ideal $I = f(M)$ is finitely generated. Therefore by Proposition 6.2 we obtain an element $r \in Ann(JI)\setminus Ann(I)$ where J is the maximal ideal of R. Now it is routine to check that Rrf is a simple submodule of M^*, and hence M^* is an essential extension of $Soc(_RM^*) = Ann_{M^*}(J)$ and $Soc(_RM^*)$ is the dual of the finitely generated semisimple module M/JM; consequently $Soc(_RM^*)$ is also finitely generated. Thus M^* is finitely cogenerated by Proposition 1.9.

PROPOSITION 6.4. Let I be any ideal of a linearly compact subdirectly irreducible ring R. If $Ann(I) \subseteq Ann(a)$ for some $a \in R$, then there is a finitely generated ideal K contained in I with $Ann(K) \subseteq Ann(a)$.

PROOF. The case $a = 0$ is trivial. So let $a \neq 0$. Since $Ann(I) = \bigcap_{r \in I} Ann(r)$ and $R/Ann(a) \cong Ra$ is finitely cogenerated, the result follows from Corollary 3.10.

PROPOSITION 6.5. Let I be any ideal of a linearly compact subdirectly irreducible ring R. If $Ann(I) \subseteq Ann(a)$ for some $a \in R$, then there is an element $b \in I$ with $Ann(a) = Ann(b)$.

PROOF. The case $a = 0$ is trivial. Therefore we assume $a \neq 0$. By Propositipn 6.4, the set of all finitely generated ideals K contained in I with $\operatorname{Ann}(K) \subseteq \operatorname{Ann}(a)$ is not empty. Consider a descending chain of finitely generated ideals $K_i \subseteq I$ satisfying $\operatorname{Ann}(K_i) \subseteq \operatorname{Ann}(a)$. For $K = \cap K_i$ we have $\operatorname{Ann}(K) = \cup \operatorname{Ann}(K_i) \subseteq \operatorname{Ann}(a)$ by Proposition 6.1. Applying Propostion 6.4 we obtain a finitely generated ideal $T \subseteq K$ with $\operatorname{Ann}(T) \subseteq \operatorname{Ann}(a)$. This implies by Zorn's lemma the existence of a minimal finitely generated ideal K with $\operatorname{Ann}(K) \subseteq \operatorname{Ann}(a)$. Note that if L is any ideal of R contained in K with $\operatorname{Ann}(L) \subseteq \operatorname{Ann}(a)$ then, by Proposition 6.4, there exists a finitely generated ideal L_1 contained in L such that $\operatorname{Ann}(L_1) \subseteq \operatorname{Ann}(a)$. Thus, by the minimality of K we get $L_1 = L = K$ and hence K is minimal among the ideals L contained in I such that $\operatorname{Ann}(L) \subseteq \operatorname{Ann}(a)$. For the maximal ideal J of R we have $K \neq JK$ by Nakayama's Lemma. Let

$$f_1 \colon \operatorname{Ann}(JK)/\operatorname{Ann}(K) \longrightarrow (K/JK)^*$$

be the homomorphism via

$$f_1(x + \operatorname{Ann}(K))(y + JK) = xy, \quad x \in \operatorname{Ann}(JK),\ y \in K.$$

It is easy to see that f_1 is a monomorphism. Let N be the image of $\operatorname{Ann}(JK)/\operatorname{Ann}(K)$ under f_1 in $M = (K/JK)^*$. Since K/JK is a finite direct sum of simple modules and R is subdirectly irreducible, M is also a finite direct sum of simple modules. On the other hand $N \neq 0$ by the minimality of K and the monomorphism of f_1. Assume now indirectly that $N \neq M$. This ensures the existence of a non-zero submodule $\bar{P}$ properly contained in K/JK such that for each non-zero element $n \in N$ there is an element $\bar{y} \in \bar{P}$ with $n\bar{y} \neq 0$. For the inverse image P of $\bar{P}$ under the canonical epimorphism $K \longrightarrow K/JK$ we obtain from the above consideration that P is properly contained in K and for each $x \in \operatorname{Ann}(JK)\setminus\operatorname{Ann}(K)$ there is an element $y \in P$ with $xy \neq 0$. Consequently we have $\operatorname{Ann}(P) = \operatorname{Ann}(K) \subseteq \operatorname{Ann}(a)$ which contradicts to the minimality of K. Therefore $N = M$ holds and thus f_1 is an isomorphism. By Proposition 5.3, K^* is an essential extension of M. For $A = K + Ra$ we have the equality

$\text{Ann}(A) = \text{Ann}(K) \cap \text{Ann}(a) = \text{Ann}(K)$. Consider now the following mapping

$$f_2\colon \text{Ann}(a)/\text{Ann}(K) \longrightarrow (A/Ra)^*,$$

defined by

$$f_2(x + \text{Ann}(K))(y + Ra) = xy, \quad x \in \text{Ann}(a),\ y \in A.$$

One can see that f_2 is a monomorphism. By the isomorphism $A/Ra = (K + Ra)/Ra \cong K/(K \cap Ra)$ the module $\text{Ann}(a)/\text{Ann}(K)$ can be considered as a submodule of $(K/(K \cap Ra))^*$ and hence of K^*. Since K^* is finitely cogenerated, the above consideration implies the non-zero intersection
$T = (\text{Ann}(a)/\text{Ann}(K)) \cap (\text{Ann}(JK)/\text{Ann}(K))$ in the case $\text{Ann}(a) \neq \text{Ann}(K)$. T is contained properly in $\text{Soc}(K^*) = \text{Ann}(JK)/\text{Ann}(K)$. Otherwise $\text{Ann}(a) \supseteq \text{Ann}(JK)$ which contradicts to the minimality of K. We claim that $T = 0$. Assume indirectly $\text{Ann}(a) \neq \text{Ann}(K)$. Then the isomorphism f_1 ensures the nontrivial direct decomposition $K/JK = \bar{P} \oplus \bar{Q}$ such that the dual $\bar{Q}$ is T and $\bar{P}$ is the annihilator (or in other words the orthogonal complement) of T. This shows for the inverse image P of $\bar{P}$ under the epimorphism $K \longrightarrow K/JK$ that P is properly contained in K and $\text{Ann}(P)/\text{Ann}(K) = T$. Thus $\text{Ann}(P) \subseteq \text{Ann}(a)$ which contradicts to the minimality of K. Thus we obtain the equality $\text{Ann}(a) = \text{Ann}(K)$. Since $Ra \cong R/\text{Ann}(a)$, we know $\text{Ann}(a)$ is sheltered and then $K = Rb$ for some b by the minimality of K and hence our proof is complete.

PROPOSITION 6.6. Let R be a linearly compact subdirectly irreducible ring. If $\text{Ann}(a) = \text{Ann}(b)$ for two elements $a,b \in R$, then there is an element $r \in R$ with $a = rb$.

PROOF. The case $a = 0$ is trivial. Therefore we assume $a \neq 0$. Since R is subdirectly irreducible, there is an element $c \in R$ such that ac generates the smallest non-zero ideal H of R. By assumption we get also $H = Rbc$. Therefore there is an element $r \in R$ with $rbc = ac$ and hence $Ann(a) \subset Ann(a-rb)$. Consider now the set of all ideals I_i of R satisfying $Ann(a) \subset I_i$ and $(a-r_ib)I_i = 0$ for some $r_i \in R$. The above discussion implies that this set is non-empty. This set is partially ordered by inclusion. If $I_1 \leq I_2$, then we have for the corresponding elements r_1,r_2 the equalities $(a-r_2b)I_1 = (a-r_1b)I_1 = 0$ and hence $(r_1-r_2)bI_1 = 0$, i.e., $r_1-r_2 \in Ann(bI_1)$. Consider now an ascending chain $\{I_i\}$. The above consideartion implies that the family $\{r_i, Ann(bI_i\}_i$ is finitely solvable. Therefore by the linear compactness of R this family is solvable, so there is an $r \in R$ with $(r-r_i)bI_i = 0$ for all i. Let $I = \cup I_i$ we have $I \geq I_i$ for all indices i and $(a-rb)I = 0$. Hence by Zorn's Lemma there is a maximal ideal I with with $(a-rb)I = 0$ for some $r \in R$. We claim $I = R$. Assume $I \neq R$. Since R is a local ring, $I \leq J$ where J is the maximal ideal of R. Since $Ann(a) \subset Ann(a-br) = I$, by Proposition 6.5 there is an element s such that $Ann(sa) = Ann(a-br)$. Now $Ann(a) \subset Ann(a-br)$, so $s \in J$. Since $I \neq R$, $a-br \neq 0$, and so $sa \neq 0$. Let $H = R(a-rb)d = Rsad$ for some d and let $(a-rb)d = tsad$ for some t. It follows that $(a-rb)d \neq 0$ but $(a-rb-tsa)d = 0$, and hence $Ann(a-rb) \subset Ann(a-rb-tsa) = Ann((1-ts)a-rb)$. Since $s \in J$, we have $ts \in J$ and then $1-ts$ is a unit of R. Let $v \in R$ with $v(1-ts) = 1$ we obtain $Ann((1-ts)a-rb) = Ann(a-vrb)$ and hence for $u = vr$ we get $I \subset Ann(a-ub)$ which contradicts to the maximality of I. Hence $I = R$ and thus $a = rb$ and our proof is complete.

PROPOSITION 6.7. Let R be a commutative subdirectly irreducible ring. The following assertions are equivalent:

(1) R is linearly compact;

(2) ${}_RR$ is injective.

PROOF. (2) ==> (1). Since ${}_RR$ is injective and R is subdirectly irreducible, ${}_RR$ is a cogenerator by Proposition

1.13. By the commutativity of R we conclude that ${}_RR_R$ induces a duality by Theorem 2.4. And then R is linearly compact by Muller's theorem.

(1) ==> (2). We show that R has the double annihilator property, i.e., $I = Ann(Ann(I))$ for each ideal I of R. The inclusion $I \subseteq Ann(Ann(I))$ is trivial. For the converse, let $a \in Ann(Ann(I))$, then we get $Ann(I) \subseteq Ann(a)$. By Propositions 6.5 and 6.6 we obtain $a \in I$. Consequently $I = Ann(Ann(I))$. Now by Proposition 5.3, ${}_RR$ is F-injective, and then ${}_RR$ is injective by Proposition 5.2.

Now we are ready to prove the main result in this section.

THEOREM 6.8 (Ánh [90]). Every linearly compact commutative ring R has a self-duality.

PROOF. Since every linearly compact commutative ring is a finite direct product of local rings, one can assume that R is local by Corollary 4.7. We shall show that the minimal injective cogenerator ${}_RU = E(R/J(R))$ is linearly compact. Let $0 \neq u \in U$, we have $Ru \cong R/Ann_R(u)$, where $Ann_R(u) = \{r \in R \mid ru = 0\}$. So $R/Ann_R(u)$ is a linearly compact subdirectly irreducible ring. By Proposition 6.7, Ru is an injective $R/Ann_R(u)$-module, and then it is easy to see that $Ru = Ann_U(Ann_R(u))$. Therefore $Ru = uS$ and this implies that S being a subdirect product of commutative rings $R/Ann_R(u) \cong S/Ann_S(u)$ $(u \in U)$, is a commutative ring. Hence by the linear compactness, we know ${}_RR$ is U-reflexive (Theorem 4.1) and then

$$S \subseteq End(U_S) = R \subseteq End({}_RU) = S,$$

i.e., R = S. So ${}_RU_R$ is a faithfully balanced bimodule such that both ${}_RU$ and U_R are injective cogenerators, i.e., ${}_RU_R$ defines a self-duality.

CHAPTER 2
MORITA DUALITY AND RING EXTENSIONS

In this chapter, we consider Morita duality and ring extensions. In Section 7, two examples show that if $S \geq R$ is a finite extension then there is no connection between the duality of R and that of S. According to Vámos [77] and Ánh [90], a ring which is linearly compact over its center has a self-dulaity. Haack [82] proved that a finite subdirect product of rings with dualities has itself a duality. In Section 8, we study finite triangular extensions. Lemonnier [84] has proved that a finite triangular extension $S \geq R$ over a ring R with a duality has itself a duality. Although the converse is open at this time, some partial results are included. Finite normalizing extensions are considered in Section 9. If R has a self-duality and $S \geq R$ is a finite normalizing extension, Mano [84] showed that, under some additional conditions, S has a self-duality. This is a generalization of a result of Fuller and Haack [81]. Kraemer [90] has proved that a finite normalizing extension over a division ring has self-duality. Trivial extensions $R \propto M$ of a ring R by an R-bimodule M are introduced in Section 10. This is due to Müller [69]. Faith's theorem [79] on cogenerator rings (or PF-rings) for trivial extensions is also presented.

SECTION 7. SOME BASIC FACTS

If $S \geq R$ is a ring extension, then an S-module ${}_SM$ is automatically an R-module ${}_RM$. It follows from the definition that if ${}_RM$ is finitely cogenerated (resp., linearly compact) then so is ${}_SM$.

PROPOSITION 7.1. Let $S \geq R$ be a ring extension. If ${}_RE$ is an injective module (or a cogenerator), then so is ${}_S\mathrm{Hom}_R(S, E)$.

PROOF. For any S-module ${}_SM$, by Adjoint Isomorphism we have

$$\mathrm{Hom}_S({}_SM, \mathrm{Hom}_R(S,E)) \cong \mathrm{Hom}_R(S \otimes M, E) \cong \mathrm{Hom}_R(M, E).$$

The results follow easily.

LEMMA 7.2. Let $S \geq R$ be a ring extension such that ${}_RS$ is linearly compact. If ${}_RE$ is a finitely cogenerated injective cogenerator, then ${}_S\mathrm{Hom}_R(S,E)$ is an essential extension of its socle.

PROOF. It suffices to show that Sf is an finitely cogenerated S-module for each $f \in \mathrm{Hom}_R(S,E)$. Since ${}_RS$ is linearly compact, so is ${}_RR$. Notice that $\mathrm{Lan}_S(f) = \{s \in S \mid sf = 0\}$ is the largest left ideal of S contained in $\mathrm{Ker}(f)$. Let $\{S_i\}_{i \in I}$ be an inverse system of left ideals of S containing $\mathrm{Lan}_S(f)$. If $\bigcap_{i \in I} S_i = \mathrm{Lan}_S(f) \subseteq \mathrm{Ker}(f)$ then, since ${}_RS$ is linearly compact and $S/\mathrm{Ker}(f) \cong \mathrm{Im}(f) \leq {}_RE$ is finitely cogenerated, it follows from Corollary 3.10 that $S_j \subseteq \mathrm{Ker}(f)$ for some $j \in I$. But then $S_j = \mathrm{Lan}_S(f)$ since $\mathrm{Lan}_S(f)$ is the largest left ideal of S contained in $\mathrm{Ker}(f)$. Accordingly, $S/\mathrm{Lan}_S(f) \cong Sf$ is finitely cogenerated by Proposition 1.5.

PROPOSITION 7.3 (Müller [84]). Let $S \geq R$ be a ring extension. If ${}_RS$ is linearly compact and R has a Morita duality induced by ${}_RE$ such that ${}_R\mathrm{Hom}_R(S,E)$ is linearly compact, then S has a Morita duality induced by ${}_S\mathrm{Hom}_R(S,E)$.

PROOF. Recall (Theorem 4.5) that ${}_RE$ defines a Morita duality if and only if ${}_RR$ is linearly compact and ${}_RE$ is a linearly compact and finitely cogenerated injective cogenerator.

Let $S^* = \mathrm{Hom}_R(S,E)$. Since ${}_RS$ and ${}_RS^*$ are linearly compact, so are ${}_SS$ and ${}_SS^*$. Thus $\mathrm{Soc}_S(S^*)$, being a submodule of the linearly compact module ${}_SS^*$, is linearly compact. Hence $\mathrm{Soc}({}_SS^*)$ is finitely generated since it is linearly compact and semisimple. It follows from Lemma 7.2 that ${}_SS^*$ is finitely cogenerated. Now ${}_SS^*$ is an injective cogenerator by Proposition 7.1. Therefore S has a Morita duality induced by ${}_SS^* = {}_S\mathrm{Hom}_R(S,E)$.

A ring extension $S \geqq R$ is called a <u>finite extension</u> in case both ${}_RS$ and S_R are finitely generated R-modules. The following two examples show that, for a finite extension $S \geqq R$, there is no connection between the duality of R and that of S. For this reason, we study finite triangular extensions in the next section. We shall see that the latter bahaves much better than the former.

EXAMPLE 7.4. Let $R = \begin{bmatrix} \mathbb{Q} & \mathbb{Q} & \mathbb{Q} \\ 0 & \mathbb{Z} & \mathbb{Q} \\ 0 & 0 & \mathbb{Q} \end{bmatrix}$ and $S = M_3(\mathbb{Q})$, the ring of 3 by 3 matrices over $\mathbb{Q}$. Then $S \geqq R$ is a finite extension. In fact $S = \sum_{i=1}^{4} Rs_i = \sum_{i=1}^{4} s_iR$, where $s_1 = 1_S$, $s_2 = \begin{bmatrix} 0&0&0 \\ 1&0&0 \\ 0&0&0 \end{bmatrix}$, $s_3 = \begin{bmatrix} 0&0&0 \\ 0&0&0 \\ 1&0&0 \end{bmatrix}$, and $s_4 = \begin{bmatrix} 0&0&0 \\ 0&0&0 \\ 0&1&0 \end{bmatrix}$. Now S is semisimple hence has a Morita duality which can be induced by ${}_SS_S$. But R is not even semilocal, so R is not semiperfect hence R does not have a Morita duality.

EXAMPLE 7.5. The artinian ring $S = \begin{bmatrix} D & D \\ 0 & C \end{bmatrix}$ in Remark 2.9 does not have a Morita duality, but S is a finite extension over the semisimple ring $R = \begin{bmatrix} D & 0 \\ 0 & C \end{bmatrix}$ since both ${}_DD$ and D_C are finitely generated.

We conclude this section by establishing two more interesting theorems.

THEOREM 7.6 (Vámos [77]). Let R be a commutative ring with a self-duality induced by ${}_RE_R$. If S is a ring whose center $C(S) \supseteq R$ and S is a linearly compact R-module (e.g., S_R is finitely generated) then the S-bimodule ${}_S\mathrm{Hom}_R(S,E)_S$ defines a self-dulaity.

PROOF. Since R is a commutative ring and ${}_RE_R$ defines a self-duality, we may assume that $rx = xr$ for each $r \in R$ and $x \in E$. Hence $\mathrm{Hom}_R({}_RS,{}_RE) = \mathrm{Hom}_R(S_R,E_R)$, and we denote this S-bimodule by ${}_SU_S$. By Proposition 7.1, both ${}_SU$ and U_S are injective cogenerators. Using Theorem 1.21, we have a right R-isomorphism

$$\mathrm{Hom}_S({}_SU,{}_SU) = \mathrm{Hom}_S({}_SU,\ {}_S\mathrm{Hom}_R(S,E))$$
$$\cong \mathrm{Hom}_R(S\otimes U,E) \cong \mathrm{Hom}_R(U,E) = \mathrm{Hom}_R(\mathrm{Hom}_R(S,E),E)) \cong S_R$$

where the last isomophism is the evaluation map since S_R is E-reflexive. It is routine to check that each of the above isomorphisms is also a right S-homomorphism. Hence $S_S \cong \mathrm{Hom}_S({}_SU,{}_SU)_S$ and S_S is ${}_SU_S$-reflexive. By the symmetry, ${}_SS$ is also ${}_SU_S$-reflexive. Therefore ${}_SU_S$ is a faithful balanced S-bimodule (Theorem 1.20) which defines a duality (Theorem 2.4).

COROLLARY 7.7. If S is a ring which is linearly compact as a $C(S)$-module, then S has a self-duality.

Let $R_1, R_2, \ldots, R_n$ be a finite number of rings. A ring R is called a <u>subdirect product</u> of R_i's in case there is a ring monomorphism $f : R \longrightarrow R_1 \times \ldots \times R_n$ such that each $p_if : R \longrightarrow R_i$ is a ring epimorphism, where $p_i : R_1 \times \ldots \times R_n \longrightarrow R_i$ is the i-th projection. In this case, let $I_i = \mathrm{Ker}(p_if)$, which is an ideal of R, then $R_i \cong R/I_i$ and

$\cap_{i=1}^n I_i = 0$. Conversely, if R is a ring with ideals $I_1,\dots,I_n$ and $\cap_{i=1}^n I_i = 0$, then R is a subdirect product of $R/I_1,\dots,$ R/I_n, where the ring monomorphism $f : R \longrightarrow (R/I_1) \times\dots\times (R/I_n)$ is the canonical homomorphism $r \longmapsto (r+I_1,\dots,r+I_n)$. The following result is a generalization of Corollary 4.7.

THEOREM 7.8 (Haack [82]). Let R be a subdirect product of the rings $R_1,\dots,R_n$. Then R has a duality if and only if each R_i has a duality.

PROOF. Let $R_i = R/I_i$ for some ideal I_i of R and then $\cap_{i=1}^n I_i = 0$. The implication (==>) follows from Corollary 2.5.

(<==). Each R/I_i is linearly compact as a left R-module. Since linearly compact modules are perserved by submodules, factor modules and finite direct product, we see that ${}_RR$ is linearly compact. Let ${}_RE$ be a finitely cogenerated injective cogenerator. Being finitely cogenerated, $\mathrm{Ran}_E(I_i)$ is linearly compact as an R/I_i-module, hence as an R-module. Since ${}_RE$ is injective, using induction and an idea in the proof of Proposition 5.1 we see that

$$\Sigma_{i=1}^n \mathrm{Ran}_E(I_i) = \mathrm{Ran}_E(\cap_{i=1}^n I_i) = \mathrm{Ran}_E(0) = E.$$

Hence E is a factor module of the linearly compact module $\oplus_{i=1}^n \mathrm{Ran}_E(I_i)$, and so ${}_RE$ is linearly compact. It follows that ${}_RE$ defines a duality.

SECTION 8. FINITE TRIANGULAR EXTENSIONS

Finite triangular extensions arise from Azumaya's exact rings that we shall study in Chapter 4. Unlike finite extensions (Examples 7.4 and 7.5), a finite triangular extension over a ring with a duality has itself a duality. In fact, any intermediate ring has such a duality (Theorem 8.4). The converse will be considered (Theorems 8.5 and 8.6), too.

LEMMA 8.1. Let ${}_SX$ and ${}_RE$ be modules. Let $f : X \longrightarrow E$ be an additive injection such that f carries every S-submodule of X to an R-submodule of E. If ${}_RE$ is finitely cogenerated (resp., linearly compact), then so is ${}_SX$.

PROOF. Let ${}_RE$ be finitely cogenerated. If $\{X_i\}_{i\in I}$ is a family of submodules of ${}_SX$ with $\bigcap_{i\in I} X_i = 0$, then each $f(X_i)$ is an R-submodule of E by hypothesis and also $\bigcap_{i\in I} f(X_i) = f(\bigcap_{i\in I} X_i) = 0$, where the equalitys hold since f is an additive injection. Hence there is some finite subset $F \subseteq I$ such that $f(\bigcap_{i\in F} X_i) = \bigcap_{i\in F} f(X_i) = 0$, since ${}_RE$ is fintely cogenerated. It follows that $\bigcap_{i\in F} X_i = 0$.

Now let ${}_RE$ be linearly compact. If $\{y_i, X_i\}_{i\in I}$ is a finitely solvable family of ${}_SX$, then $\{f(y_i), f(X_i)\}_{i\in I}$ is a finitely solvable family of ${}_RE$. Hence there exists an $a \in E$ such that $a-f(y_i) \in f(X_i)$ for all $i \in I$. Let $a-f(y_i) = f(x_i)$ for some $x_i \in X_i$, then $a = f(y_i+x_i) \in f(X)$. So $a = f(x)$ for some $x \in X$, and $x = y_i+x_i$ since f is injective. Hence $x-y_i = x_i \in X_i$ for all $i \in I$. We conclude that ${}_SX$ is linearly compact.

Dually, one can prove the following

LEMMA 8.2. Let ${}_SX$ and ${}_RE$ be modules. Let $f : X \longrightarrow E$ be an additive surjection such that $f^{-1}(L)$ is an S-submodule of X for each R-submodule L of E. If ${}_SX$ is finitely generated (resp., linearly compactc), then so is ${}_RE$.

A ring extension $S \geq R$ is called a <u>finite triangular extension</u> in case there are a finite number of elements $s_1,\dots,s_n$

in S such that $S = \sum_{i=1}^{n} Rs_i$ and $\sum_{i=1}^{j} Rs_i = \sum_{i=1}^{j} s_iR$ $(j = 1,\dots,n)$. The latter n equalities hold if and only if for each $r \in R$, there are upper triangular matrices $[a_{ij}]$ and $[b_{ij}]$ over R such that $rs_j = \sum_{i=1}^{j} s_ia_{ij}$ and $s_jr = \sum_{i=1}^{j} b_{ij}s_i$ $(j = 1,\dots,n)$.

PROPOSITION 8.3. Let $S \geq R$ be a finite triangular extension. Then

(1) If ${}_RE$ is finitely cogenerated (resp., linearly compact), then $\mathrm{Hom}_R(S, E)$ is finitely cogenerated (resp., linearly compact) as an R- and hence as an S-module.

(2) If ${}_SV$ is finitely cogenerated, then V is finitely cogenerated as an R-module.

PROOF. Let $S = \sum_{i=1}^{n} Rs_i$ be such an extension. We use induction on n to show that ${}_S\mathrm{Hom}_R(S,E)$ is finitely cogenerated (resp., linearly compact). Set $S_1 = \sum_{i=1}^{n-1} Rs_i$, and consider the map $f: \mathrm{Hom}_R(S/S_1, E) \longrightarrow E$ given by $f(g) = g(s_n+S_1)$. It is easy to see that f is an additive injection. Let $G \leq {}_R\mathrm{Hom}_R(S/S_1, E)$ and $f(g) \in f(G)$ for some $g \in G$. Since $S/S_1 = R(s_n+S_1) = (s_n+S_1)R$, for each $r \in R$ we have $r(s_n+S_1) = (s_n+S_1)r'$ for some $r' \in R$. Then

$$\begin{aligned} r(f(g)) &= r(g(s_n+S_1) = g(rs_n+S_1) = g(s_nr'+S_1) \\ &= (r'g)(s_n+S_1) = f(r'g) \in f(G), \end{aligned}$$

since $r'g \in G$. This shows that f carries each R-submodule of $\mathrm{Hom}_R(S/S_1, E)$ to an R-submodule of E. So by Lemma 8.1, ${}_R\mathrm{Hom}_R(S/S_1, E)$ is finitely cogenerated (resp., linearly compact). Now from the exact sequence of R-bimodules

$$0 \longrightarrow S_1 \longrightarrow S \longrightarrow S/S_1 \longrightarrow 0$$

we get an exact sequence of left R-modules

$$0 \longrightarrow {}_R\mathrm{Hom}_R(S/S_1,\ E) \longrightarrow {}_R\mathrm{Hom}_R(S,E) \longrightarrow {}_R\mathrm{Hom}_R(S_1,\ E).$$

Since the last term is finitely cogenerated (resp., linearly compact) by induction hypothesis, so is ${}_R\mathrm{Hom}_R(S,E)$ by Propositions 1.8 and 3.3.

(2) Let $\{T_i \mid i \in I\}$ be an irredundant set of representatives of the simple left R-modules, and let ${}_RU = \oplus_I E(T_i)$. Since ${}_RS$ is finitely generated, $\mathrm{Hom}_R(S,\ U) \cong \oplus_I \mathrm{Hom}_R(S,\ E(T_i))$ canonically. Now ${}_S\mathrm{Hom}_R(S,U)$ is a cogenerator (Proposition 7.1), since ${}_RU$ is. And ${}_SV$ is finitely cogenerated, it can be embedded in a direct sum of finitely many modules of the form $\oplus_F \mathrm{Hom}_R(S,\ E(T_i))$, where $F \subseteq I$ is a finite subset. It follows that ${}_RV$ is finitely cogenerated by (1).

THEOREM 8.4. Let $S \geqq T \geqq R$ be ring extensions such that $S \geqq R$ is a finite triangular extension. If R has a Morita duality induced by ${}_RE$, then T has a Morita duality induced by ${}_T\mathrm{Hom}_R(T,E)$. In particular, S has a Morita duality induced by ${}_S\mathrm{Hom}_R(S,E)$.

PROOF. Since ${}_RR$ is linearly compact and ${}_RS$ is finitely generated, ${}_RS$ is linearly compact and hence ${}_RT$ is linearly compact. Using the injectivity of ${}_RE$, the exact sequence of R-bimodules

$$0 \longrightarrow T \longrightarrow S \longrightarrow S/T \longrightarrow 0$$

induces an exact sequence of left R-modules

$$0 \longrightarrow {}_R\mathrm{Hom}_R(S/T,E) \longrightarrow {}_R\mathrm{Hom}_R(S,E) \longrightarrow {}_R\mathrm{Hom}_R(T,E) \longrightarrow 0.$$

Now ${}_R\mathrm{Hom}_R(S,E)$ is linearly compact by Proposition 8.3, hence ${}_R\mathrm{Hom}_R(T,E)$ is linearly compact. We have proved that ${}_RT$ and ${}_R\mathrm{Hom}_R(T,E)$ are linearly compact, and hence ${}_T\mathrm{Hom}_R(T,E)$ induces a Morita duality by Propostion 7.3.

The following question arises naturally from the above theorem: If $S \geqq R$ is a finite triangular extension and S has a Morita duality, does R have a duality? We are unable to settle this problem, but we do have two partial solutions.

THEOREM 8.5. Let $S \geqq R$ be a finite triangular extension and S have a Morita duality induced by $_SV$. If both $_RS$ and S_R are progenerators, then R has a duality induced by $_RV$.

THEOREM 8.6. Let $S \geqq R$ be a finite triangular extension. If S is a left artinian ring with a duality, then so is R.

To prove these two theorems, we need some preparations.

LEMMA 8.7. Let $S \geqq R$ be a ring extension with $_RS$ a generator. Let $_RM$ and N_R be R-modules. Then $_RM$ (resp. N_R) is linearly compact whenever $_S\mathrm{Hom}_R(S,M)$ (resp. $(N\otimes_R S)_S$) is linearly compact. In particular, if S_S is linearly compact, then so is R_R.

PROOF. Since $_RS$ is a generator and $_RR$ is finitely generated, there is a natural number n with an epimorphism $_RS^n \longrightarrow {}_RR$. Let $(s_1,\dots,s_n) \longmapsto 1$ and let $_RS \longrightarrow {}_RS^n$ via $s \longmapsto (ss_1,\dots,ss_n)$, then the composite map $p : {}_RS \longrightarrow {}_RR$ has the property that $p(r) = r$ for all $r \in R$.

(1) Assume that $_S\mathrm{Hom}_R(S,M)$ is linearly compact. For mappings

$$f : M \longrightarrow \mathrm{Hom}_R(S,M), \quad f(m)(s) = p(s)m,$$

and

$$g : \mathrm{Hom}_R(S,M) \longrightarrow M, \quad g(h) = h(1),$$

we have $gf = 1_M$. If $\{m_i, M_i\}_i$ is a finitely solvable famity in $_RM$, so is $\{f(m_i), M_i'\}_i$ in $\mathrm{Hom}_R(S,M)$ where $M_i' = \mathrm{Hom}_R(S,M_i)$. There is some $h \in \mathrm{Hom}_R(S,M)$ such that $h-f(m_i) \in M_i'$ for all i, which implies that $h(1)-m_i = g(h)-gf(m_i) \in g(M_i') = M_i$ for all

i. It follows that ${}_RM$ is linearly compact.

(2) Now we assume that $(N\otimes_R S)_S$ is linearly compact. Define a map $f : N\otimes S \longrightarrow N$ via $n\otimes s \longmapsto np(s)$. It is an additive epimorphism and for any $L \le N_R$ we have $f^{-1}(L) = L\otimes S \le (N\otimes S)_S$, so by Lemma 8.2 N_R is linearly compact.

The following two assertions are straightfowardly verified, so we omit the proofs.

LEMMA 8.8. Let $S \geq R$ be a ring extension and ${}_RE$ a left injective R-module. If ${}_R\mathrm{Hom}_R(S, E)$ has finite length, then ${}_RE$ has finite length.

LEMMA 8.9. Let $S = \sum_{i=1}^{n} Rs_i \geq R$ be a finite triangular extension. Let ${}_SM$ be an S-module with an R-submodule L. For each $s \in S$ let $s^{-1}L = \{m \in M \mid sm \in L\}$. Set $M_0(L) = M$, and $M_j(L) = \bigcap_{i=1}^{j} s_i^{-1}L$ ($j = 1,\dots,n$). Then each $M_j(L) \leqq {}_RM$ ($j = 0,1, \dots,n$) and $M_n(L) \leqq {}_SM$. For each j, the natural group monomorphism $M_j(L)/M_{j+1}(L) \longrightarrow M/L$ defined by $m+M_{j+1}(L) \longmapsto s_{j+1}m+L$ induces a lattice embedding $\mathfrak{L}(M_j(L)/M_{j+1}(L)) \longrightarrow \mathfrak{L}(M/L)$ of the lattices of R-submodules.

PROPOSITION 8.10. Let $S = \sum_{i=1}^{n} Rs_i \geq R$ be a finite triangular extension. Let ${}_SM$ be an S-module. If ${}_SM$ has finite length, then ${}_RM$ also has finite length.

PROOF. It clearly suffices to consider the case of a simple S-module ${}_SM$. Then ${}_SM = Sm$ is cyclic and ${}_RM = \sum_{i=1}^{n} Rs_im$ is finitely generated. Let L be a maximal R-submodule of ${}_RM$. Using the notations in Lemma 8.9, we have $M_n(L) = 0$ since ${}_SM$ is simple and $M_n(L)$ is an S-submodule of $L \neq M$. We have a chain of R-submodules of M:

$$0 = M_n(L) \leqq M_{n-1}(L) \leqq \ldots \leqq M_1(L) \leqq M_0(L) = M.$$

By Lemma 8.9, each $M_j(L)/M_{j+1}(L)$ can be embedded to M/L, in a way which preserves submodules, by the map $m+M_{j+1}(L) \longmapsto s_{j+1}m+L$. Hence each $M_j(L)/M_{j+1}(L)$ is simple or zero, so the length of ${}_RM$ is no greater than n.

PROOF of Theorem 8.5. Since ${}_SS$ is linearly compact, so is ${}_RR$ by Lemma 8.7. Since ${}_SV$ is finitely cogenerated, by Proposition 8.3, ${}_RV$ and ${}_S\mathrm{Hom}_R(S,V)$ are finitely cogenerated. Hence ${}_S\mathrm{Hom}_R(S,V)$ is linearly compact since ${}_SV$ is a linearly compact cogenerator. It follows that ${}_RV$ is linearly compact by Lemma 8.7 again. For any left R-module ${}_RM$, we have

$$\mathrm{Hom}_R(M,V) \cong \mathrm{Hom}_R(M, \mathrm{Hom}_S(S, V) \cong \mathrm{Hom}_S(S\otimes_R M, V),$$

and the projectivity of S_R implies that ${}_RV$ is injective. Let ${}_RT$ be any simple module. Since S_R is a generator, we have $S\otimes T \neq 0$. Thus, since ${}_SV$ is a cogenerator, $\mathrm{Hom}_S(S\otimes T, V) \neq 0$ and so $\mathrm{Hom}_R(T,V) \neq 0$, which implies that ${}_RV$ is a cogenerator. Hence ${}_RV$ defines a Morita duality by Theorem 4.5.

PROOF of Theorem 8.6. Using Proposition 8.10, ${}_RS$ has finite length. Then R is a left artinian ring, so R has a finitely cogenerated injective cogenerator ${}_RE$. It follows from Propositions 7.1 and 8.3 that ${}_S\mathrm{Hom}_R(S, E)$ is a finitely cogenerated injective cogenerator which has finite length since S is a left artinian ring with a duality (Corollary 4.4). Then ${}_R\mathrm{Hom}_R(S, E)$ also has finite length by Proposition 8.10 again. It follows from Lemma 8.8 that ${}_RE$ has finite length. Consequently, R has a duality induced by ${}_RE$.

SECTION 9. FINITE NORMALIZING EXTENSIONS

A finite triangular extension $S = \sum_{i=1}^{n} Rs_i \geq R$ is called a <u>finite normalizing extension</u> in case $Rs_i = s_iR$ $(i = 1,\ldots,n)$. If G is a finite group, then the group ring RG is such an extension over R. An immediate corollary of Theorem 8.4 is the following

THEOREM 9.1 (Kitamura [81]) Let $S \geq R$ be a finite normalizing extension. If R has a Morita duality induced by ${}_RE$, then S has a Morita duality induced by ${}_S\mathrm{Hom}_R(S,E)$.

In this section we consider finite normalizing extensions and self-duality. We shall prove two main results that are Theorem 9.2 (Mano) and Theorem 9.11 (Kraemer).

THEOREM 9.2 (Mano [84a]). Let $S = \sum_{i=1}^{n} Rs_i \geq R$ be a finite normalizing extension satisfying the followung:

(1) ${}_RS$ is a free R-module with basis $s_1,\ldots,s_n$ such that each s_i centralizes the elements of R, i.e., $rs_i = s_ir$ for all $r \in R$; and

(2) R has self-duality induced by ${}_RE_R$ such that $b_{ijp}x = xb_{ijp}$ for all $x \in E$ and all i, j, p; where we put $s_is_j = \sum_{p=1}^{n} b_{ijp}s_p$.

Then S has self-duality induced by ${}_S\mathrm{Hom}_R(S,E)$.

Before we prove Mano's theorem, we like to mention two corollaries.

COROLLARY 9.3. Every finite dimensional algebra R over a field satisfies the conditions in Theorem 9.1, hence R has self-duality.

If R is a ring and G is a finite semigroup (with an

identity) then the <u>semigroup ring</u> RG has free basis G which centralizes the elements of R, and the multiplication is defined as

$$(\Sigma_G r_g g)(\Sigma_G r'_g g) = \Sigma_G(\sum_{hk=g} r_h r'_k)g.$$

We may regard R as a subring of RG under the embedding $R \hookrightarrow RG$ via $r \longmapsto r1_G$. Hence the following result follows from Mano's Theorem.

COROLLARY 9.4 (Fuller and Haack [81]). Let R be a ring with self-duality. If G is a finite semigroup then the semigroup ring RG has self-duality.

Fuller and Haack [81] actually proved more: If R is a ring and an R-module $_RE$ defines a Morita duality between R and $S = \mathrm{End}(_RE)$, then $_{RG}\mathrm{Hom}_R(RG, E)$ defines a Morita duality between RG and SG. Hence, if R has self-duality, then so does RG.

PROOF. Let $V = {_{RG}\mathrm{Hom}_R(RG, E)}$. By Theorem 9.1, $_{RG}V$ defines a duality, so it remains to prove that $SG \cong \mathrm{End}(_{RG}V)$ as rings. We will show that the composition of the natural abelian group isomorphism is in addtion a ring homomorphism.

The map

$$f_1\colon SG \longrightarrow \mathrm{Hom}_R(E^{(G)}, E)$$

defines via

$$f_1(\Sigma s_g g)\colon (e_g) \longmapsto \Sigma e_g s_g \quad (\Sigma s_g g \in SG,\ (e_g) \in E^{(G)})$$

is an abelian group isomorphism. Define

$$f_2\colon V = \mathrm{Hom}_R(RG_R, E) \longrightarrow {_RE^{(G)}}$$

via

$$f_2\colon v \longmapsto (v(g)) \quad (g \in G \le RG,\ v \in V).$$

Then f_2 is a left R-isomorphism, since G centralizes R and $RG = \oplus Rg$ is a free left R-module. Let $f_2^* = \mathrm{Hom}_R(f_2, E)$ be the

E-dual map of f_2,

$$f_2^*: \mathrm{Hom}_R({}_RE^{(G)}, {}_RE) \longrightarrow \mathrm{Hom}_R({}_RV, {}_RE).$$

Define

$$f_3: (RG \otimes_{RG} V) \longrightarrow V$$

via

$$f_3: t\otimes v \longmapsto tv \quad (t \in Rg,\ v \in V);$$

then f_3 is a left R-isomorphism and

$$f_3^* = \mathrm{Hom}_R(f_3, {}_RE) : \mathrm{Hom}_R({}_RV, {}_RE) \longrightarrow \mathrm{Hom}_R(RG\otimes_{RG}V, {}_RE)$$

is also an isomorphism. Finally, define

$$f_4: \mathrm{Hom}_R({}_R(RG\otimes_{RG}V), {}_RE) \longrightarrow \mathrm{Hom}_{RG}({}_{RG}V, \mathrm{Hom}_R({}_RRG_{RG}, {}_RE)$$

via

$$[f_4(h)](v): t \longrightarrow h(t\otimes v)$$

where $h \in \mathrm{Hom}_R({}_R(RG\otimes_{RG}V), {}_RE)$, $v \in V$, $t \in RG$. f_4 is an isomorphism by adjointness (Theorem 1.21). It follows that

$$f = f_4f_3^*f_2^*f_1: SG \longrightarrow \mathrm{End}({}_{RG}V)$$

is an abelian group isomorphism. Now we show that f preserves multiplication. First, notice that for $\Sigma\, s_g g \in SG$, $v \in V$, and $t \in RG$,

$$(v)(f(\Sigma\, s_g g)): t \longmapsto \Sigma v(gt)s_g.$$

Let $\Sigma\, s'_g g \in SG$. Then

$$\begin{aligned}
[(v)f((\Sigma\, s_g g)(\Sigma\, s'_g g))](t) &= [(v)f(\sum_g(\sum_{hk=g} s_h s'_k)g)](t) \\
&= \sum_g v(gt)(\sum_{hk=g} s_h s'_k) = \sum_g \sum_{hk=g} v(gt)s_h s'_k \\
&= \sum_k \sum_h v(hkt)s_h s'_k = \sum_k (\sum_h v(hkt)s_h)s'_k \\
&= \sum_k \Big\{ [(v)f(\sum_h s_h h)](kt) \Big\} s'_k \\
&= \Big\{ (v)[f(\sum_h s_h h)f(\Sigma\, s'_k k)] \Big\}(t),
\end{aligned}$$

and f is multiplicative. Hence f is a ring isomorphism from

SG to $\mathrm{End}({}_{RG}V)$.

The following question was raised in Fuller [F]: Let R be a ring and G a finite (semi-) group. If RG has self-duality, does R have self-duality?

Now we prove Theorem 9.2 and make the assumptions of Theorem 9.2 until the end of Lemma 9.10.

LEMMA 9.5. Let $1 = \sum_{i=1}^{n} a_i s_i$ and let $C(R)$ be the center of R. Then

(1) b_{ijp}, $a_i \in C(R)$, for all i,j and p.

(2) $\sum_p b_{ijp} b_{pkm} = \sum_p b_{jkp} b_{ipm}$, for all i,j,k and m.

(3) $\sum_p a_p b_{pji} = \delta_{ij} = \sum_p a_p b_{jpi}$, for all i and j.

PROOF. (1) Since $\sum_p b_{ijp} r s_p = (\sum_p b_{jkp} s_p) r = (s_i s_j) r = r(s_i s_j) = r(\sum_p b_{ijp} s_p) = \sum_p r b_{ijp} s_p$ for all $r \in R$, and $\{s_1, \dots, s_n\}$ is a basis of ${}_R S$, we have $b_{ijp} r = r b_{ijp}$. Next, since $\sum_i a_i r s_i = (\sum_i a_i s_i) r = 1r = r1 = r(\sum_i a_i s_i) = \sum_i r a_i s_i$ for all $r \in R$, we get $a_i r = r a_i$ for all $r \in R$ and i.

(2) $(s_i s_j) s_k = (\sum_p b_{ijp} s_p) s_k = \sum_p b_{ijp} \sum_m b_{pkm} s_m$. On the other hand, $s_i (s_j s_k) = s_i (\sum_p b_{jkp} s_p) = (\sum_p b_{jkp} s_i s_p) = (\sum_p b_{jkp} \sum_m b_{ipm} s_m)$. So (2) follows.

(3) Since $\sum_i (\sum_p a_p b_{pji}) s_i = \sum_p a_p (\sum_i b_{pji} s_i) = \sum_p a_p s_p s_j) = (\sum_p a_p s_p) s_j = s_j = \sum_i \delta_{ij} s_i$ for all j, we have $\sum_p a_p b_{pji} = \delta_{ij}$ for all i and j.

From the hypothesis of Theorem 9.2, R has self-duality

induced by ${}_R E_R$, hence ${}_R E$ and E_R are linearly compact and finitely cogenerated injective cogenerators, $\mathrm{End}({}_R E) \cong R$, and $\mathrm{End}(E_R) \cong R$. Put ${}_R W_R = ({}_R E_R)^{(n)}$ and denote the elements of W as row vectors $[e_t]$. For each $\sum_i r_i s_i \in S$ and $[e_t] \in W$, we define

$$(\sum_i r_i s_i) * [e_t] = [\sum_p \sum_i r_i b_{tip} e_p].$$

LEMMA 9.6. With the multiplication "*", W becomes a left S- right R-bimodule. Moreover, left R-module structure of $({}_R E)^{(n)}$ coincides with the multiplication "*", i.e., $(\sum_i r a_i s_i) * [e_t] = [r e_t]$ for all $r \in R$ and $[e_t] \in W$.

PROOF. We shall only prove $s * (s' * [e_t]) = (ss') * [e_t]$ for all $s, s' \in S$ and $[e_t] \in W$. Let $s = \sum_j r_j s_j$ and $s' = \sum_k r'_k s_k$, then

$$\begin{aligned} s * (s' * [e_t]) &= (\sum_j r_j s_j) * ((\sum_k r'_k s_k) * [e_t]) \\ &= (\sum_j r_j s_j) * [\sum_q \sum_k r'_k b_{tkq} e_q] = [\sum_p \sum_j \sum_q \sum_k r_j b_{tjp} r'_k b_{pkq} e_q]. \end{aligned}$$

On the other hand,

$$\begin{aligned} (ss') * [e_t] &= (\sum_j r_j s_j)(\sum_k r'_k s_k) * [e_t] \\ &= (\sum_p \sum_k \sum_j r_j r'_k b_{jkp} s_p) * [e_t] = [\sum_q \sum_p \sum_k \sum_j r_j r'_k b_{jkp} b_{tpq} e_q]. \end{aligned}$$

Using Lemma 9.5(1)(2) we have $s * (s' * [e_t]) = (ss') * [e_t]$.

From now on, we will denote the multiplication omitting "*".

LEMMA 9.7. $F : {}_S\mathrm{Hom}_R({}_R S_S \, , {}_R E_R) \ni f \longmapsto [f(s_t)] \in {}_S W_R$ is a bimodule isomorphism.

PROOF. Since ${}_R S_R = ({}_R R_R)^{(n)}$, it is easy to show that F

is an R-bimodule isomorphism. Let $\sum_k r_k s_k \in S$ and $f \in \mathrm{Hom}_R({}_RS, {}_RE)$. Then

$$(\sum_k r_k s_k)[f(s_t)] = [\sum_p \sum_k r_k b_{tkp} f(s_p)] = [f(\sum_p \sum_k r_k b_{tkp} s_p]$$
$$= [f(\sum_k r_k \sum_p b_{tkp} s_p)] = [f(\sum_k r_k s_t s_k)]$$
$$= [f(s_t \sum_k r_k s_k)] = [(\sum_k r_k s_k f)(s_t)].$$

Hence $(\sum_k r_k s_k)F(f) = F(\sum_k r_k s_k f)$, which shows that F is an S-homomorphism.

The following result follows from Theorem 9.1 and Lemma 9.7.

PROPOSITION 9.8. ${}_SW$ is a linealy compact and finitely cogenerated injective cogenerator that defines a Morita duality between S and $\mathrm{End}({}_SW)$.

To show that $S \cong \mathrm{End}({}_SW)$, we now proceed to compute $\mathrm{End}({}_SW)$. Since $\mathrm{End}({}_RW) = \mathrm{End}({}_RE^{(n)}) = \mathbb{M}_n(\mathrm{End}({}_RE)) = \mathbb{M}_n(R)$, we have $R \subseteq \mathrm{End}({}_SW) \subseteq \mathrm{End}({}_RW) = \mathbb{M}_n(R)$. We will denote the elements in $\mathbb{M}_n(R)$ of the form $\left[r_{pq} \right]$.

LEMMA 9.9. $G : S \ni \sum_i r_i s_i \longmapsto \left[\sum_i r_i b_{iqp} \right] \in \mathbb{M}_n(R)$ is a ring monomorphism.

PROOF. Clearly G is an additive homomorphism. We first prove that G is monic.

$$\sum_i r_i s_i \in \mathrm{Ker}(G) \implies \sum_i r_i b_{iqp} = 0 \text{ for all } p \text{ and } q,$$
$$\implies \sum_p \sum_i r_i b_{iqp} s_p = 0 \text{ for all } q,$$
$$\implies \sum_i r_i s_i s_q = 0 \text{ for all } q,$$

$$\Longrightarrow \sum_q \sum_i r_i s_i s_q a_q = 0,$$

$$\Longrightarrow \sum_i r_i s_i = 0.$$

Hence G is monic. Next we prove that G is a ring homomorphism. Let $\sum_i r_i s_i$, $\sum_j r_j' s_j \in S$. Then

$$G(\sum_i r_i s_i)G(\sum_j r_j' s_j) = \left[\sum_i r_i b_{iqp} \right]\left[\sum_j r_j' b_{jqp} \right]$$

$$= \left[\sum_i \sum_j r_i r_j' \sum_t b_{itp} b_{jqt} \right].$$

On the other hand,

$$G((\sum_i r_i s_i)(\sum_j r_j' s_j)) = G(\sum_i \sum_j r_i r_j' \sum_t b_{ijt} s_t) =$$

$$= \left[\sum_i \sum_j r_i r_j' \sum_t b_{ijt} b_{tqp} \right].$$

Thus $G(\sum_i r_i s_i)G(\sum_j r_j' s_j) = G((\sum_i r_i s_i)(\sum_j r_j' s_j))$ by Lemma 9.5(2).

The following commutative diagram follows from the above lemma.

$$
\begin{array}{ccc}
S & \xrightarrow{\;\;G\;\;} & \mathbb{M}_n(R) \\
\uparrow & & \uparrow \\
R & \hookrightarrow & \mathrm{End}(_S W)
\end{array}
$$

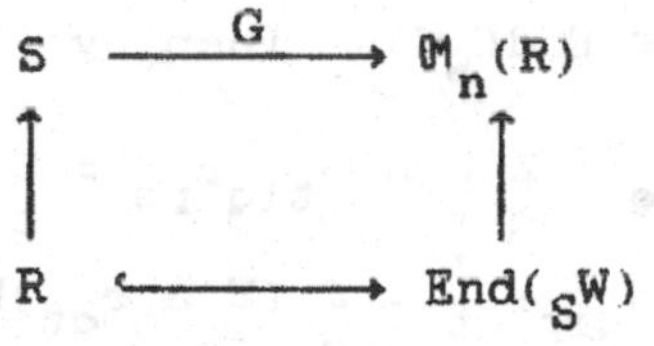

The proof that $S \cong \mathrm{End}(_S W)$ (hence the proof of Theorem 9.2) will be completed after we establish the following

LEMMA 9.10. Under the assumptions of Theorem 9.2, we have

(1) Let $\left[r_{pq} \right] \in \mathbb{M}_n(R)$. Then $\left[r_{pq} \right] \in \mathrm{End}(_S W)$ if and only if $\sum_q b_{tiq} r_{hq} = \sum_q b_{qih} r_{qt}$ for all i, h and t.

(2) $\mathrm{End}(_S W) = \sum_{k=1}^{n} R\left[b_{kqp} \right] = G(S)$.

PROOF. (1) Let $[e_t] \in W$. Then

$s_i([e_t]\left(r_{pq}\right)) = s_i[\sum_h e_h r_{ht}] = [\sum_h \sum_q b_{tiq} e_h r_{hq}] = [\sum_h \sum_q e_h b_{tiq} r_{hq}]$

for all i. On the other hand,

$$(s_i[e_t])\left(r_{pq}\right) = [\sum_h b_{tih} e_h]\left(r_{pq}\right)$$

$$= [\sum_h \sum_q b_{qih} e_h r_{qt}] = [\sum_h \sum_q e_h b_{qih} r_{qt}] \text{ for all } i.$$

Thus we have $\left(r_{pq}\right) \in \mathrm{End}({}_S W)$ if and only if

$\sum_h \sum_q e_h b_{tiq} r_{hq} = \sum_h \sum_q e_h b_{qih} r_{qt}$ for all $[e_t] \in W$ and i, t. (#)

Suppose $\left(r_{pq}\right) \in \mathrm{End}({}_S W)$. Let $e \in E$ and fix t. Put $[e_t] = [\delta_{pt} e]$. Then by (#), $e\sum_q b_{tiq} r_{hq} = e\sum_q b_{qih} r_{qt}$ for all $e \in E$. Therefore $\sum_q b_{tiq} r_{hq} = \sum_q b_{qih} r_{qt}$. Conversely, if $\sum_q b_{tiq} r_{hq} = \sum_q b_{qih} r_{qt}$ for all i and h, then it is easy to see that $\left(r_{pq}\right) \in \mathrm{End}({}_S W)$ from (#).

(2) Let $\left(r_{pq}\right) \in \mathrm{End}({}_S W)$. Then by (1),

$$r_{pi} = \sum_q \delta_{qi} r_{pq} = \sum_q \sum_t a_t b_{tiq} r_{pq} = \sum_t a_t \sum_q b_{tiq} r_{pq}$$

$$= \sum_t a_t \sum_q b_{qip} r_{qt} = \sum_q (\sum_t a_t r_{qt}) b_{qip}$$

for all p and i. Put $c_q = \sum_t a_t r_{qt}$. Then we have $r_{pq} = \sum_k c_k b_{kqp}$ for all p and q. Thus $\left(r_{pq}\right) = \sum_k c_k \left(b_{kqp}\right) \in \sum_k R\left(b_{kqp}\right)$. On the other hand, one can easily check that $\sum_k R\left(b_{kqp}\right) \subseteq \mathrm{End}({}_S W)$. Thus we have proved (2).

The other main result in this section is due to Kraemer.

THEOREM 9.11 (Kraemer [90]). If S is a finite normalizing extension over a division ring D, then S has self-duality induced by ${}_S\mathrm{Hom}_D({}_DS,{}_DD)$.

PROOF. Let $S = \sum_{i=1}^{n} Ds_i \geq D$ be a finite normalizing extension over a division ring D. Then $\{s_1,\ldots,s_n\}$ contains a basis of the D-vector space ${}_DR$, hence we can assume without restriction that $S = \bigoplus_{i=1}^{n} Ds_i$. Let ${}_SU = {}_S\mathrm{Hom}_D({}_DS,{}_DD)$ that induces a duality by Theorem 9.1.

Since $Ds_i = s_iD$ and D is a division ring, there are automorphisms $f_i\colon D \cong D$, $d \longmapsto d'$ with $s_id = d's_i$. Let $u_j\colon {}_DS \longrightarrow {}_DD$, $\Sigma_i\, d_is_i \longmapsto d_j$ be the j-th projection; we claim that $U = \oplus_j\, Du_j$ with $du_j = u_jf_j(d)$, in particular $Du_j = u_jD$ ($d \in D$, $j = 1,\ldots,n$): Evidently $U = \oplus_j\, Du_j$, and for $d \in D$ and $j,k = 1,\ldots,n$ we calculate

$$(du_j)(s_k) = u_j(s_kd) = \delta_{jk}f_k(d) = (u_jf_j(d))(s_k)$$

(with the Kronecker symbol δ_{jk}), hence $du_j = u_jf_j(d)$.

If $T = \mathrm{End}({}_SU)$, then ${}_SU_T$ is a bimodule, and from the Adjoint isomorphism we infer that $F : T \longrightarrow \mathrm{Hom}_D({}_DU,{}_DD)$, $t' \longmapsto (u \longmapsto (ut')(1))$ is a T-D-biisomorphism with the inverse $G(t)(u)(s) = t(su)$ ($t \in \mathrm{Hom}_D({}_DU,{}_DD)$, $u \in U$, $s \in S$). Let $t_k\colon {}_DU \longrightarrow {}_DD$, $\Sigma_j\, d_ju_j \longmapsto d_k$ be the k-th projection; then $\mathrm{Hom}_D({}_DU,{}_DD) = \oplus_k\, Dt_k$ with $t_kd = f_k(d)t_k$, in particular $Dt_k = t_kD$ ($d \in D$, $k = 1,\ldots,n$), as is shown with the same arguments as above. Denote $t_k' = G(t_k) \in T$; then $T = \oplus_k\, Dt_k'$ with $t_k'd = f_k(d)t_k'$ ($d \in D$, $k = 1,\ldots,n$).

There are $a_{ijk}, a'_{ijk}, a^*_{ijk}, a^{**}_{ijk} \in D$ ($i,j,k = 1,\ldots,n$) such that $s_is_j = \Sigma_m\, a_{ijm}s_m$, $s_ju_k = \Sigma_m\, a'_{kjm}u_m$, $u_kt_i' = \Sigma_m\, a^*_{ikm}u_m$, $t_i't_j' = \Sigma_m\, a^{**}_{ijm}t_m'$. In order to prove that R has self-duality it is now sufficient to show that $a_{ijk} = a^{**}_{ijk}$ ($i,j,k = 1,\ldots,n$). Then $H\colon T \longrightarrow R$, $\Sigma_i\, d_it_i' \longmapsto \Sigma_i\, d_is_i$ is a ring isomorphism and the statement follows since ${}_R U_T$ defines a duality.

For $i,j,k = 1,\dots,n$ we evaluate as follows.

$(s_j u_k)(s_i) = u_k(s_i s_j) = u_k(\Sigma_m a_{ijm} s_m) = a_{ijk}$ and $(s_j u_k)(s_i) = (\Sigma_m a'_{kjm} u_m)(s_i) = f_i(a'_{kji})$;

$(u_k t'_i)(s_j) = (s_j u_k t'_i)(1) = F(t'_i)(s_j u_k) = t_i(\Sigma_m a'_{kjm} u_m) = a'_{kji}$ and $(u_k t'_i)(s_j) = (\Sigma_m a^*_{ikm} u_m)(s_j) = f_j(a^*_{ikj})$;

$(u_k t'_i t'_j)(1) = F(t'_j)(u_k t'_i) = t_j(\Sigma_m a^*_{ikm} u_m) = a^*_{ikj}$ and $(u_k t'_i t'_j)(1) = (u_k \Sigma_m a^{**}_{ijm} t'_m)(1) = F(\Sigma_m a^{**}_{ijm} t'_m)(u_k) = \Sigma_m (t_m f_k^{-1}(a^{**}_{ijm}))(u_k) = f_k^{-1}(a^{**}_{ijk})$. Hence $a_{ijk} = f_i f_j f_k^{-1}(a^{**}_{ijk})$ $(i,j,k = 1,\dots,n)$.

On the other hand $s_i s_j d = (\Sigma_k a_{ijk} s_k)d = \Sigma_k a_{ijk} f_k(d) s_k$ and $s_i s_j d = f_i f_j(d) s_i s_j = f_i f_j(d)(\Sigma_k a_{ijk} s_k)$, thus, comparing coefficients, $a_{ijk} f_k(d) = f_i f_j(d) a_{ijk}$ $(d \in D,\ i,j,k = 1,\dots,n)$. If we put $d = f_k^{-1}(a_{ijk})$, then $a^2_{ijk} = f_i f_j f_k^{-1}(a_{ijk}) a_{ijk}$, hence $a_{ijk} = f_i f_j f_k^{-1}(a_{ijk})$ for $i,j,k = 1,\dots,n$ (we can cancel if $a_{ijk} \neq 0$). Then $a_{ijk} = a^{**}_{ijk}$ $(i,j,k = 1,\dots,n)$, thus R has self-duality.

Finite normalizing extensions of division rings occur quite frequently. The most common examples are the finite dimensional algebras over fields. In conclusion we mention the crossed semigroup rings over division rings as discussed in Kraemer [90]: Let D be a division ring, G a finite semigroup (with the identity e), and $R = \oplus_{g\in G} R_g$ a strongly G-graded ring with $R_e = D$ (that is R is a ring with $R_g R_h = R_{gh}$ for all $g,h \in G$). If, for all $g \in G$, $R_g = D\bar{g} = \bar{g}D \neq 0$ for some $\bar{g} \in R_g$, then R is a _crossed semigroup ring of_ G _over_ D. Particular examples of such rings are the _twisted semigroup rings of_ G _over_ D

(that is $d\bar{g} = \bar{g}d$ for all $g \in G$ and $d \in D$), and the <u>skew semigroup ring of</u> G <u>over</u> D (that is $\bar{g}\bar{h} = \overline{gh}$ for all g, $h \in G$).

SECTION 10. TRIVIAL EXTENSIONS

A ring A with ideal M and subring R such that $M^2 = 0$ and $A = R \oplus M$ is called a <u>trivial extension</u> of R by M, and is denoted by $A = R \propto M$. If ${}_RM_R$ is a bimodule then one constructs such an ring extension from $R \times M$ with multiplication

$$(r_1,m_1)(r_2,m_2) = (r_1r_2,\ r_1m_2 + m_1r_2).$$

The following theorem is due to Müller [69].

THEOREM 10.1 (Müller [69]). Let ${}_RE_S$ define a duality and let ${}_RM_R$ be an R-bimodule. Then $R \propto M$ has a finitely cogenerated injective cogenerator $\mathrm{Hom}_R(M,E) \propto E$, with multiplication given by $(r,m)(f,x) = (rf,\ rx+f(m))$, and its endomorphism ring is $S \propto \mathrm{Hom}_R(\mathrm{Hom}_R(M,E),E)$. We have a duality between these rings if and only if both ${}_RM$ and ${}_R\mathrm{Hom}_R(M,E)$ are linearly compact.

PROOF. Since M is a nilpotent ideal in $R \propto M = A$ hence in the radical, so $J(A) = J(R) \oplus M$. Consequently, R and A have the same simple modules. Now ${}_A\mathrm{Hom}_R(A,E)$ is an injective cogenerator (Proposition 7.1) containning

$${}_A\mathrm{Hom}_R(A,\mathrm{Soc}({}_RE)) \geq {}_A\mathrm{Hom}_A(A,\mathrm{Soc}({}_RE)) \cong \mathrm{Soc}({}_RE).$$

Hence to show ${}_A\mathrm{Hom}_R(A,E)$ is finitely cogenerated, it suffices to show that it is essential over ${}_A\mathrm{Hom}_A(A,\mathrm{Soc}({}_RE))$. In fact, for any $0 \neq f \in {}_A\mathrm{Hom}_R(A,E)$, either $f(0,m_0) \neq 0$ for some $m_0 \in M$, then $0 \neq f(0,r_0m_0) \in \mathrm{Soc}({}_RE)$ for some $r_0 \in R$ since ${}_RE$ is essential over its socle, and $(0,r_0m_0)\cdot f$ is non-zero and in ${}_A\mathrm{Hom}_A(A,\mathrm{Soc}({}_RE))$; or $f(0,m) = 0$ for all $m \in M$, then

$f(1,0) \neq 0$ and $0 \neq f(r_1,0) \in \mathrm{Soc}({}_R E)$ for some $r_1 \in R$, therefore $(r_1,0)\cdot f$ is non-zero and in ${}_A\mathrm{Hom}_A(A,\mathrm{Soc}({}_R E))$. Now

$$\mathrm{Hom}_R(A,E) \cong \mathrm{Hom}_R(M\oplus R,E) \cong \mathrm{Hom}_R(M,E)\oplus E$$

as abelian group, and it is easily checked that $R \propto M$ operates as indicated. Next

$$\begin{aligned}\mathrm{End}({}_A\mathrm{Hom}_R(A,E)) &\cong \mathrm{Hom}_R(A\otimes_A\mathrm{Hom}_R(A,E),E)\\ &\cong \mathrm{Hom}_R(\mathrm{Hom}_R(R\propto M,E),E) \cong S \propto \mathrm{Hom}_R(\mathrm{Hom}_R(M,E),E))\end{aligned}$$

and we check that the multiplication is given by $(s,h)(s',h')=(ss',sh' + hs')$; hence we have a trivial ring extension of S by the S-bimodule $H = \mathrm{Hom}_R(\mathrm{Hom}_R(M,E),E))$ of the same type as $R \propto M$, and it operates on $\mathrm{Hom}_R(M,E) \propto E$ by $(f,x)(s,h) = (fs,\ fh + xs)$.

If $R \propto M$ has a duality, then it has a duality induced by ${}_{R\propto M}\mathrm{Hom}_R(M,E) \propto E$. Then ${}_{R\propto M}\mathrm{Hom}_R(M,E) \propto E$ is linearly compact and $R \propto M$ is left lineaely compact. If $\{f_i,\ F_i\}_{i\in I}$ is a finitely solvable family of ${}_R\mathrm{Hom}_R(M,E)$, then $\{(f_i,0),\ F_i\oplus E\}_{i\in I}$ is a finitely solvable family of ${}_{R\propto M}\mathrm{Hom}_R(M,E) \propto E$. Let (f,x) be a solution for $\{(f_i,0),\ F_i\oplus E\}_{i\in I}$. Then f is a solution for $\{f_i,\ F_i\}_{i\in I}$, and so ${}_R\mathrm{Hom}_R(M,E)$ is linearly compact. Similarly, using the linear compactness of ${}_{R\propto M}(R \propto M)$, one verifies that ${}_R M$ is linearly compact.

Now assuming that both ${}_R M$ and ${}_R\mathrm{Hom}_R(M,E)$ are linearly compact, we show that $R \propto M$ has a duality induced by ${}_{R\propto M}(\mathrm{Hom}_R(M,E) \propto E)$. Since ${}_R R$ and ${}_R M$ are linearly compact, so is ${}_R(R \propto M)$, hence ${}_{R\propto M}(R \propto M)$ is linearly compact. Similarly, the linear compactness of ${}_R\mathrm{Hom}_R(M,E)$ and ${}_R E$ implies that $\mathrm{Hom}_R(M,E) \propto E$ is linearly compact as R- hence as $R \propto M$-module. It follows that the finitely cogenerated injective cogenerator ${}_{R\propto M}(\mathrm{Hom}_R(M,E) \propto E)$ defines a duality.

COROLLARY 10.2. Let ${}_R E_R$ define a self-duality and let ${}_R M_R$ be an R-bimodule. If both ${}_R M$ and ${}_R\mathrm{Hom}_R(M,E)$ are linearly compact, and there is an isomorphism f of R and an f-semilinear R-R-bimodule isomorphism

$$g : {}_R\mathrm{Hom}_R({}_RM, {}_RE)_R \cong {}_R\mathrm{Hom}_R(M_R, E_R),$$

i.e., $g(r_1hr_2) = f(r_1)g(h)\, f(r_2)$, then $A = R \propto M$ has self-duality induced by $\mathrm{Hom}_R({}_RM, {}_RE) \propto E$.

PROOF. Let $H = {}_R\mathrm{Hom}_R(\mathrm{Hom}_R({}_RM, {}_RE), {}_RE)_R$. It follows from the theorem that $\mathrm{Hom}_R({}_RM, {}_RE) \propto E$ defines a duality between left $(R \propto M)$-modules and right $(R \propto H)$-modules. Using the isomorphism g we have $\alpha : H \cong \mathrm{Hom}_R(\mathrm{Hom}_R(M_R, E_R), {}_RE) \cong M$, where the composite isomorphism is an f-semilinear R-R-bimodule isomorphism. Then the map

$$R \propto H \longrightarrow R \propto M \text{ via } (r,h) \longmapsto (f(r), \alpha(h))$$

is a ring isomorphism.

Let R and S be rings and ${}_RM_S$ a bimodule. Let $\begin{bmatrix} R & M \\ 0 & S \end{bmatrix}$ denote the formal triangular matrix ring as defined in Section 2. Then we have

COROLLARY 10.3. Let R and S have duality induced by ${}_RE$ and ${}_SF$, respectively. Then the ring $A = \begin{bmatrix} R & M \\ 0 & S \end{bmatrix}$ has a duality if and only if both ${}_RM$ and ${}_S\mathrm{Hom}_R(M,E)$ are linearly compact. In this case ${}_AG = \begin{bmatrix} E & 0 \\ \mathrm{Hom}_R(M,E) & F \end{bmatrix}$ defines a duality between left A- and right $B = \begin{bmatrix} \mathrm{End}({}_RE) & \mathrm{Hom}_S(\mathrm{Hom}_R(M,E),F)) \\ 0 & \mathrm{End}({}_SF) \end{bmatrix}$-modules.

PROOF. If $(r,s) \in R \oplus S$ and $m \in M$, let $(r,s)m = rm$ and $m(r,s) = ms$. Then M becomes an $R \oplus S$-bimodule and $A \cong (R \oplus S) \propto M$ via $\begin{bmatrix} r & m \\ 0 & s \end{bmatrix} \longmapsto ((r,s),m)$. The results now follow from the theorem.

The following result follows from Corollaries 10.2 and 10.3.

COROLLARY 10.4. Let R and S have self-duality induced by ${}_RE_R$ and ${}_SF_S$, respectively. Suppose that both ${}_RM$ and ${}_S\mathrm{Hom}_R(M,E)$ are linearly compact, and there are ring isomorphisms f_1 of R and f_2 of S such that there is an f_1-f_2-similinear R-S-bimodule isomorphism $g : {}_S\mathrm{Hom}_R(M,E)_R \cong {}_S\mathrm{Hom}_S(M,F)_R$, i.e., $g(shr) = f_2(s)g(h)f_1(r)$. Then $A = \begin{bmatrix} R & M \\ 0 & S \end{bmatrix}$ has a self-duality.

Using Corollary 10.3, we easily see that the ring $\begin{bmatrix} D & D \\ 0 & C \end{bmatrix}$ in Remark 2.9 does not have a duality. In fact, we have the following more general result.

COROLLARY 10.5. Let D be a division ring with a division subring C. Then the ring $A = \begin{bmatrix} D & D \\ 0 & C \end{bmatrix}$ has a duality if and only if ${}_CD$ is finitely generated. In this case, ${}_AG = \begin{bmatrix} D & 0 \\ D & C \end{bmatrix}$ defines a duality between left A- and right $B = \begin{bmatrix} D & \mathrm{Hom}_C({}_CD, {}_CC) \\ 0 & C \end{bmatrix}$-modules.

PROOF. Since ${}_DD_D$ and ${}_CC_C$ deifne self-duality and ${}_C\mathrm{Hom}_D({}_DD_C, {}_DD) \cong {}_CD$, the results follow from Corollary 10.3.

Recall that a ring R is a cogenerator ring if ${}_RR_R$ induces a self-duality (see Section 5). Next we present Faith's Theorem that the trivial extension $R \propto E$ is a cogenerator ring if ${}_RE_R$ defines a self-duality of R. Then we give a non-noetherian cogenerator ring as promised in Section 5.

We begin with the main lemma used in the proof of Theorem 10.7.

LEMMA 10.6. Let S be a ring, let E be an ideal such that $E = \mathrm{Ran}_S(E)$, let $R = S/E$. Then E is canonically an R-module. If

(i) E is injective as a (canonical) left R-module, and

(ii) $R \cong \mathrm{End}({}_R E)$ canonically,

then ${}_S S$ is injective.

Conversely, if ${}_S S$ is injective, then for any ideal A, the right annihilator $\mathrm{Ran}_S(A)$ is an injective left S/A-module, and $\mathrm{End}({}_{S/A}\mathrm{Ran}_S(A)) \cong S/\mathrm{Ran}_S(\mathrm{Lan}_S(\mathrm{Ran}_S(A)))$ canonically. Thus, in this case, any ideal E satisfying $E = \mathrm{Ran}_S(E)$ satisfies (i) and (ii).

PROOF. Let $F = E({}_S S)$, and let $F_1 = \mathrm{Ran}_F(E)$. Then, F_1 is a left R-module, and $E = \mathrm{Ran}_S(E)$ is an injective left R-module by the assumption (i). Since every R-submodule of F_1 is a S-submodule, then F_1 is an essential extension of $F_1 \cap S = E$ as a S-module, hence as an R-module, so the injectivity of ${}_R E$ implies that $F_1 = \mathrm{Ran}_F(E) = E$. Thus, if $y \in F$, then $Ey \subseteq \mathrm{Ran}_F(E) = E$, so y induces an endomorphism $b \in B = \mathrm{End}({}_S E) = \mathrm{End}({}_R E)$. Now every $s \in S$ induces an endomorphism $s_m \in \mathrm{End}({}_R E)$ via right multiplication; hence $R = S/\mathrm{Ran}_S(E) = S/E$ embeds in B canonically. Since $R \cong B$ canonically by the hypothesis (ii), there exists $s \in S$ such that

$$xy = b(x) = xs_m = xs, \text{ for each } x \in E,$$

so

$$x(y - s) = 0, \text{ for each } x \in E;$$

hence

$$y - s = t \in \mathrm{Ran}_F(E) = E \subseteq S.$$

Therefore, $y = s + t \in S$, for each $y \in F$, proving that $F = {}_S S$ is injective.

Conversely, let ${}_S S$ be injective and A be an ideal of S. By Proposition 1.10, ${}_{S/A}\mathrm{Ran}_S(A)$ is an injective S/A-module. Now each $b \in \mathrm{End}({}_S A)$ is induced by an element $s \in S$; hence $S/\mathrm{Ran}_S(A) \cong \mathrm{End}({}_S A)$. Also,
$S/\mathrm{Lan}_S(\mathrm{Ran}_S(A)) \cong \mathrm{End}({}_S\mathrm{Ran}_S(A)) \cong \mathrm{End}({}_{S/A}\mathrm{Ran}_S(A))$, canonically. Taking $A = E = \mathrm{Ran}_S(E)$, we have the stated properties (i) and (ii).

THEOREM 10.7 (Faith [79]). Let ${}_RE_R$ be an R-bimodule and $S = R \propto E$, the trivial extension of R by E. Then

(1) ${}_SS$ is injective if and only if ${}_RE$ is injective and $R \cong \mathrm{End}({}_RE)$ canonically;

(2) ${}_SS$ is an injective cogenerator if and only if ${}_RE$ is an injective cogenerator and $R \cong \mathrm{End}({}_RE)$ canonically;

(3) S is a cogenerator ring if and only if the bimodule ${}_RE_R$ defines a self-duality.

PROOF. (1). We identify E with $E_1 = 0 \times E$ in S, and R with $R_1 = R \times 0$. Clearly, $R \cong R_1 \cong S/E_1$ under $r \longmapsto (r,0) \longmapsto (r,0) + E_1$, and $\mathrm{Ran}_S(E_1) = E_1$ if E_R is faithful. Thus, assuming ${}_RE$ injective and $R \cong \mathrm{End}({}_RE)$, that is assuming (i) and (ii) of the above lemma, we have ${}_SS$ is injective by the lemma. The converse also follows from this lemma.

(2) Recall that an injective module is a cogenerator if and only if it contains each simple module (Proposition 1.13(2)).

(==>). In this case, each simple left S-module ${}_ST$ embeds in ${}_SS$. Now, since J(S) contains any square-zero (or nilpotent or nil) ideal, then $J(S) \supseteq E_1$; hence $S/J(S) \cong R/J(R)$, and every simple left S-module corresponds to a simple left R-module $T_1 = R/M$. Since T embeds in ${}_SS$, then T_1 embeds in ${}_SS$. If $s \in S$ and $s = (r,x) \neq 0$ generates T, then $r = 0 \Rightarrow T \subseteq E$, and $r \neq 0 \Rightarrow$ there exists $(0,y) \neq 0 \in E$ such that $(r,x)(0,y) = (0,ry) \neq 0 \in T \cap E$; hence $T \cap E = T \subseteq E$ in both cases. This proves that each simple left R-module T_1 embeds in ${}_RE$. Since ${}_RE$ is injective by (1), this shows that ${}_RE$ is a cogenerator.

(<==). According to the correspondence of simple left S-modules and simple left R-modules as remarked in the proof of (==>), each simple left S-module T embeds in $E_1 = 0 \times E \subseteq S$. Now ${}_SS$ is injective by (1), so ${}_SS$ is a cogenerator.

(3) This follows from (2) and the left-right symmetry.

Next we present a cogenerator ring that is not noetherian, as we promised at the beginning of Section 5.

EXAMPLE 10.9. Let F be a field consider $F[[x]]$, the formal pover series ring over F. One sees that $F[[x]]$ is a commutaive noetherian ring. By Proposition 3.6, $F[[x]]$ is also linearly compact. Hence $F[[x]]$ has a self-duality by Anh's Theorem 6.8. In fact, the minimal injective cogenerator of $F[[x]]$-Mod is $F[1/x]$, the polynomial ring of F over $1/x$, where the $F[[x]]$-operation on $F[1/x]$ is induced by

$$(a_i x^i)(b_j x^{-j}) = (b_j x^{-j})(a_i x^i) = \begin{cases} 0 & i > j \\ a_i b_j x^{j-i} & \text{otherwise.} \end{cases}$$

Then $F[1/x]$ is an injective and artinian $F[[x]]$-module, and one sees that ${}_{F[[x]]}F[1/x]_{F[[x]]}$ defines a self-duality. Hence $R = F[[x]] \propto F[1/x]$ is a cogenerator ring by Faith's Theorem 10.8. Since $F[1/x]$ is not noetherian, R is a commutative cogenerator ring that is not noetherian.

In the conclusion, we discuss whether or not the conditions "finitely generated" and "finitely cogenerated" in Theorem 2.6(10) can be dropped.

Throughout let ${}_RE_S$ define a duality, then R is a semiperfect ring and assume $e_1,\dots,e_n$ is a basic set of primitive idempotents. If ${}_RP$ is projective then (see Section 1) there exist sets $A_1,\dots,A_n$ such that ${}_RP \cong Re_1^{(A_1)} \oplus \dots \oplus Re_n^{(A_n)}$. Hence if ${}_RP$ is E-reflexive and projective then each A_i is a finite set by Lemma 2.2 and so ${}_RP$ is finitely generated. It follows from Theorem 2.6(10) that the dual $\mathrm{Hom}_R(P,E)_S$ is a finitely cogenerated injective S-module. We summarize these as

PROPOSITION 10.10. Let ${}_RE_S$ define a duality. If ${}_RP$ is an E-reflexive projective R-module then ${}_RP$ is finitely generated; consequently $P^*_S = \mathrm{Hom}_R(P,E)_S$ is a finitely cogenerated injective S-module.

The above result says that the implication (==>) of Theorem 2.6(10) is still true if the condition "finitely generated" is dropped. But we shall see that for an E-reflexive injective R-module $_RU$, its dual $\mathrm{Hom}_R(U,E)_S$ need not be projective. The first such example was given in Osofsky [66], and we shall give an easier one here.

Let F be a field and $R = F[[x]] \propto F[1/x]$ as in Example 10.9. If we let $F((x))$ be the quotient field of $F[[x]]$, then $F((x))$ is the injective envelope of the commutative domain $F[[x]]$.

LEMMA 10.11. (1) $F((x))/F[[x]]$ is an artinian $F[[x]]$-module; and (2) $F((x))$ is a linearly compact $F[[x]]$-module.

PROOF. (1) Let $K = \{ q \in F((x)) \mid xq \in F[[x]] \}$. Then $q \longmapsto xq$ defines a monomorphism of $F[[x]]$-modules $K \longrightarrow F[[x]]$. Since each non-zero ideal of $F[[x]]$ is of the form $x^iF[[x]]$, K is a cyclic $F[[x]]$-module. If $0 \neq \bar{q} \in F((x))/F[[x]]$, then $\mathrm{Ann}_{F[[x]]}(\bar{q}) = \{f(x) \in F[[x]] \mid f(x)\bar{q} = 0 \} \leqq xF[[x]]$. Now $\mathrm{Ann}_{F[[x]]}(\bar{q}) \neq 0$, so there exists a n with $x^n \in \mathrm{Ann}_{F[[x]]}(\bar{q})$. Hence the simple $F[[x]]$-module $K/F[[x]] = \mathrm{Soc}(_{F[[x]]}F((x))/F[[x]])$ is essential in $F((x))/F[[x]]$. It follows that $F((x))/F[[[x]]$ can be embedded to the injective envelope of the simple $F[[x]]$-module $K/F[[x]]$, which is $F[1/x]$. Now $F[1/x]$ is artinian, so is $F((x))/F[[x]]$.

(2) Since both $F((x))/F[[x]]$ and $F[[x]]$ are linearly compact $F[[x]]$-modules, so is $F((x))$ by Proposition 3.3.

In view of Proposition 10.10, the next example shows that an E-reflexive injective R-module $_RU$ need not be finitely cogenerated and that its dual $\mathrm{Hom}_R(U,E)_S$ need not be projective, either.

EXAMPLE 10.12. By Example 10.9, $R = F[[x]] \propto F[1/x]$ is a cogenerator ring, i.e., ${}_R R_R$ defines a duality. Let ${}_R U = E({}_R R/I) = E({}_R F[[x]])$, where $I = 0 \times F[1/x]$. Hence ${}_R U = F((x))$, which is linearly compact (hence R-reflexive) by Lemma 10.11. Since $\mathrm{Soc}({}_R U) = 0$, ${}_R U$ is not finitely cogenerated. It follows from Proposition 10.10 (using the right version) that $U_R^* = \mathrm{Hom}_R(U,R_R)$ is not projective since ${}_R U^{**} \cong {}_R U$ is not finitely cogenerated.

CHAPTER 3
ARTINIAN RINGS WITH MORITA DUALITY (I)

This chapter is devoted to study left artinian rings with a duality. In Section 11, the well-known theorem of Azumaya [59] and Morita [58] about the various conditions for a left artinian ring to possess a duality is presented. Fuller's Theorem [69] is also included, which states that if Re is injective, where $e = e^2 \in R$, then ${}_{eRe}eRf_{fRf}$ defines a duality for some $f = f^2 \in R$. Quasi-Frobenius rings are also introduced and at the end of Section 11 Dieudonne's Theorem is presented, which we shall need in the next chapter. And in Section 12, we characterize artinian left duo rings via their dualities.

SECTION 11. Azumaya-Morita Theorem, Fuller Theorem, and QF-Rings

In this section, we collect the well-known Theorem of Azumaya [59] and Morita [58], and also Fuller's Theorem [69]. At the end of the section, we briefly introduce Quasi-Frobenius rings. Examples of artinian rings without a duality or with a duality but without a self-duality are given in the next section.

Recall (Corollary 4.4) that a left artinian ring R has a Morita duality if and only if every indecomposable injective left R-module is finitely generated. Jategaokar [81] has formulated the following

THEOREM 11.1. Let R be a left artinian ring with a finitely cogenerated injective cogenerator ${}_RE$ and let $S = End({}_RE)$. Then the following three conditions are equivalent:

(a) R has a duality;

(b) ${}_RE$ has finite length;

(c) S is a right artinian ring.

PROOF. (a) <==> (b) ==> (c). Corollary 4.4 and Theorem 2.8.

(c) ==> (b). Since R is left artinian, $J = J(R)$ is nilpotent, so ${}_RE = Ran_E(J^m)$ for some m. If possible, let ${}_RE$ be of infinite length. Choose smallest k such that $Ran_E(J^k)/Ran_E(J^{k-1})$ has infinite length. Since ${}_RE$ is injective, the R-endomorphism ring $\bar{S}$ of $Ran_E(J^k)$ is a factor ring of S; so, $\bar{S}$ is right artinian. Now consider the set $A = Ran_{\bar{S}}(Ran_E(J^{k-1}))$. Clearly, A is an ideal of $\bar{S}$; so $A = \Sigma_{i=1}^n f_i\bar{S}$. Notice that $Ran_E(J^{k-1}) \subseteq \cap_{i=1}^n Ker(f_i)$ and that f_i induces an R-homomorphism of the infinite dimensional semisimple R-module $Ran_E(J^k)/Ran_E(J^{k-1})$ into the finite dimensional semisimple R-module $Soc({}_RE) = Ran_E(J)$. It follows that $Ran_E(J^{k-1}) \subset (\cap_{i=1}^n Ker(f_i)) \cap Ran_E(J^k) = Lan_{(Ran_E(J^k))}(A)$. But this is false, as is easily seen by using Proposition 1.12 since ${}_RE$ is a cogenerator.

The next two results are due to Rosenberg and Zelinsky [59].

THEOREM 11.2. Let R be a semiprimary ring with $J = J(R)$. Let $J^n = 0$. If ${}_RE$ is an injective module, there is a left R-module isomorphism
$Ran_E(J^{i+1})/Ran_E(J^i) \cong Hom_R(J^i/J^{i+1}, E) = Hom_R(J^i/J^{i+1}, Soc(E))$.
Consequently,

(1) ${}_RE$ has finite length if and only if each ${}_RHom_R(J^i/J^{i+1}, Soc(E))$ has finite length, $i= 0,1,\ldots,n-1$.

(2) The condition that ${}_RHom_R(J^i/J^{i+1}, T)$ has finite length for every simple left R-module T (equivalently, for every semisimple left R-module of finite length) for $i = 1,2,\ldots$ is equivalent to the same condition postulated only for $i = 1$.

PROOF. We may identify $Ran_E(J^i)$ with $Hom_R(R/J^i, E)$ by letting correspond to x in $\mathrm{Ran}_E(J^i)$ the homomorphism sending $1+J^i$ to x. Since

$$0 \longrightarrow J^i/J^{i+1} \longrightarrow R/J^{i+1} \longrightarrow R/J^i \longrightarrow 0$$

is an exact sequence of R-bimodules, we have an exact sequence of left R-modules

$$0 \longrightarrow \mathrm{Hom}_R(R/J^i, E) \longrightarrow \mathrm{Hom}_R(R/J^{i+1}, E) \longrightarrow \mathrm{Hom}_R(J^i/J^{i+1}, E) \longrightarrow 0.$$

Making the above identification we obtain the desired isomorphism. Since $J^n = 0$, ${}_RE$ has the <u>lower Loewy series</u>

$$0 = \mathrm{Ran}_E(J^0) \leqq \mathrm{Ran}_E(J) \leqq \ldots \leqq \mathrm{Ran}_E(J^m) = E,$$

where we denote $m = L({}_RE) \leqq n$. (And the <u>upper Loewy series</u> of ${}_RE$ is $E \geqq JE \geqq \ldots \geqq J^{m-1}E \geqq J^mE = 0$.) Hence ${}_RE$ has finite length if and only if each $\mathrm{Ran}_E(J^{i+1})/\mathrm{Ran}_E(J^i)$ has finite length. Then (1) follows. For (2) we use induction on i. Consider the natural R-bimodule epimorphism $J^{i-1} \otimes_R J \longrightarrow J^i$ determined by $r_1 \otimes r_2 \longmapsto r_1 r_2$ for $r_1 \in J^{i-1}$ and $r_2 \in J$. This induces an epimorphism $(J^{i-1}/J^i) \otimes_R (J/J^2) \longrightarrow J^i/J^{i+1}$, which in turn gives an R-module monomorphism

$$\mathrm{Hom}_R(J^i/J^{i+1}, T) \longrightarrow \mathrm{Hom}_R((J^{i-1}/J^i) \otimes_R (J/J^2), T)$$
$$\cong \mathrm{Hom}_R(J^{i-1}/J^i, \mathrm{Hom}_R(J/J^2, T)),$$

where the isomorphism is by Adjoint Isomorphism. But $T' = \mathrm{Hom}_R(J/J^2, T)$ has finite length by hypothesis and is semisimple because $JT' = 0$, and so $\mathrm{Hom}_R(J^{i-1}/J^i, T')$ also has finite length by the induction hypothesis. So also do all its submodules, completing the induction.

Now we have a criterion for left artinian rings to have Morita duality.

COROLLARY 11.3. Let R be a left artinian ring with

radical J. Then the following statements are equivalent:

(1) R has a Morita duality;

(2) $\mathrm{Hom}_R(J/J^2, R/J)$ has finite length;

(3) R/J^2 has a Morita duality.

PROOF. Since each simple left R-module embeddes to R/J, ${}_RE = E(R/J)$ is a finitely cogenerated injective cogenerator.

(2) ==> (1). By Theorem 11.2, ${}_RE$ has finite length hence defines a duality.

(1) ==> (3). Corollary 2.5.

(3) ==> (2). Let $\bar{R} = R/J^2$ and $\bar{J} = J/J^2$. Now $\bar{R}$ has a duality induced by ${}_{\bar{R}}E' = E(\bar{R}/\bar{J})$, hence by Theorem 11.2 $\mathrm{Hom}_R(J/J^2, R/J) = \mathrm{Hom}_{\bar{R}}(\bar{J}, \bar{R}/\bar{J})$ has finite length.

Next, we present some basic results on indecomposable injective and projective modules over artinian rings most of which are due to Fuller [69]. First, let us write $T(M) = M/J(R)M$ for a left R-module M. A left injective module ${}_RE$ and a right projective module P_R is said to form a <u>pair</u> over a semiprimary ring R in case $E(T(Re_1)), \ldots, E(T(Re_n))$ represent all the indecomposable direct summands of ${}_RE$ where $e_1, \ldots, e_n$ is an orthogonal set of primitive idempotents such that $e_1R, \ldots, e_nR$ are (to within isomorphism) the indecomposable direct summands of P_R, and such an idempotent $e = e_1+\ldots+e_n$ is called a <u>basic idempotent</u> for ${}_RE$ and P_R.

LEMMA 11.4. Let eR_R and ${}_RE$ form a pair over a semiprimary ring R. Then

(1) $\mathrm{Ran}_E(I) = \mathrm{Ran}_E(eI)$ and $\mathrm{Lan}_{eR}(K) = \mathrm{Lan}_{eR}(KE)$ for every left ideal I and every right ideal K of R;

(2) $\mathrm{Ran}_E(eR) = 0$ and $\mathrm{Lan}_{eR}(E) = 0$;

(3) $\mathrm{Ran}_R(eR) = \mathrm{Lan}_R(E)$.

PROOF. (1) Let $X = \mathrm{Ran}_E(eI)$. It is clear that

$Ran_E(I) \le X$. Suppose $IX \ne 0$. Then, since $IX \le {}_RE$, $IX \cap Soc({}_RE) \ne 0$. This implies that IX has a simple submodule that can be embedded in $T(Re)$ and hence that $eIX \ne 0$, a contradiction. Thus $Ran_E(I) = Ran_E(eI)$. Let $L = Lan_R(KE)$. Then $eL = eR \cap L = Lan_{eR}(KE) \supseteq Lan_{eR}(K)$. Suppose $eLK \ne 0$. Then there must exist left ideals $I_1 < I_2 \leq LK$ with I_2/I_1 simple and isomorphic to a submodule of $T(Re)$. This gives a monomorphism $f : I_2/I_1 \longrightarrow E$. If $g : I_2 \longrightarrow I_2/I_1$ is the natural epimorphism we have, by the injectivity of E, that $fg(I_2) = I_2x$ for some $x \in E$. But then $LKE \supseteq I_2x = fg(I_2) \ne 0$, contrary to the definition of L. This proves that $Lan_{eR}(KE) = Lan_{eR}(K)$.

(2) This follows from (1) by taking $I = R$ and $K = R$.

(3) Let $A = Ran_R(eR)$ and $B = Lan_R(E)$. Then, using (2), $eRB = eB = Lan_{eR}(E) = 0$, so $B \le A$. Also $AE \le Ran_E(eR) = 0$, so $A \le B$.

The next results follow from Lemma 11.4(2) and Lemma 4.10.

LEMMA 11.5. Let eR_R and ${}_RE$ form a pair over a semiprimary ring R then, for each left R-module M, restriction to eM gives an abelian group isomorphism

$$Hom_R(M,E) \cong Hom_{eRe}(eM,eE).$$

Moreover,

(1) The map f defined via

$$[f(s)](ex) = s(ex),\ s \in Hom_R(E,E),\ ex \in eE,$$

is a ring isomorphism from $Hom_R(E,E)$ to $Hom_{eRe}(eE,eE)$;

(2) If $I \le K$ are ideals of R then

$$Ran_E(I)/Ran_E(K) \cong Hom_{eRe}(eK/eI,eE)$$

as left R-modules;

(3) The left eRe-module eE is injective.

If e is an idempotent of a ring R then $eJ(R)e$ is the radical of eRe, and then eRe is semiprimary if R is. So if

eR_R and $_RE$ form a pair and $e = e_1+\dots+e_n$ is a basic idempotent for eR_R and $_RE$, then $T(eRe_i) = eRe_i/eJ(R)e_i$, $i = 1,\dots,n$, are all the (non-isomorphic) simple left eRe-modules.

LEMMA 11.6. If eR_R and $_RE$ form a pair over a semiprimary ring R and $e = e_1+\dots+e_n$ is a basic idempotent for eR_R and $_RE$, then the left eRe-module $_{eRe}eE$ is an injective cogenerator over eRe. Specifically, if $_RE = \oplus E_{ij}$ with $E_{ij} \cong E(T(Re_i))$, $i=1,\dots,n$, then $_{eRe}eE = \oplus eE_{ij}$ with $eE_{ij} \cong E(T(eRe_i))$, $i=1,\dots,n$, over eRe.

PROOF. By Lemma 11.5(3) we first know that eE is injective over eRe. To complete the proof we let $_RE_{ij} = E(T(Re_i))$, $i = 1,\dots,n$, and show that $\mathrm{Soc}(_{eRe}eE_{ij}) \cong T(eRe_i)$ is simple. Since $\mathrm{Soc}(_RE_{ij})$ is simple and $e\mathrm{Soc}(E_{ij}) \supseteq e_i\mathrm{Soc}(E_{ij}) \neq 0$ it follows at once that $e\mathrm{Soc}(E_{ij})$ is simple and isomorphic to $T(eRe_i)$. Moreover, using Lemma 11.4(1),

$$\mathrm{Soc}(eE_{ij}) \subseteq \mathrm{Ran}_E(eJ(R)) \cap E_{ij} = \mathrm{Ran}_E(J(R)) \cap E_{ij} = \mathrm{Soc}(E_{ij})$$

so that

$$e\mathrm{Soc}(E_{ij}) \subseteq \mathrm{Soc}(eE_{ij}) \subseteq e\mathrm{Soc}(E_{ij}).$$

Hence $\mathrm{Soc}(eE_{ij}) = e\mathrm{Soc}(E_{ij}) \cong T(eRe_i)$ is simple.

We need the following well-known lemma (for a proof, see Anderson-Fuller [74]).

LEMMA 11.7. Let $_RM$ and $_RE_S$ be modules. Let $M^* = \mathrm{Hom}_R(M,E)$. If $_RE$ is a cogenerator then $\mathrm{Lan}_M\mathrm{Ran}_{M^*}(K)) = K$ for every $K \leq M$.

If $_RE_S$ defines a duality then the $_RE_S$-dual $(\ )^*$ takes simples to simples by Theorem 2.6. It follows that if $_RE$ is a finitely generated faithful module over a semisimple ring R, then $S = \mathrm{End}({}_RE)$ is also semisimple, ${}_RE_S$ induces a duality, and so ${}_RE_S$-dual takes simples to simples. Now we extend this to any finitely cogenerated injective cogenerator over a semilocal ring.

LEMMA 11.8. Let R be a semilocal ring and let ${}_RE$ be a finitely cogenerated injective cogenerator with $S = \mathrm{End}({}_RE)$. Then

(1) $\mathrm{Soc}({}_RE) = \mathrm{Soc}(E_S)$;

(2) The ${}_RE_S$-dual takes simples to simples.

PROOF. (1) Since $S = \mathrm{End}({}_RE)$ is a semiperfect ring by Proposition 1.19, letting $V = \mathrm{Soc}({}_RE) \leq E$, it follows from Proposition 1.11 that $J(S) = \mathrm{Ran}_S(V)$. Now we have

$$\begin{aligned} \mathrm{Soc}(E_S) &= \mathrm{Lan}_E(J(S)) && \text{(since S is semilocal)} \\ &= \mathrm{Lan}_E(\mathrm{Ran}_S(V)) \\ &= V && \text{(Lemma 11.7, since } S = E^* = \mathrm{Hom}_R(E,E)). \end{aligned}$$

(2) Using (1), write $V = \mathrm{Soc}({}_RE) = \mathrm{Soc}(E_S)$. Then, since ${}_RV$ is finitely generated and contains each copy of simple left R-module and $S/J(S)$ is naturally isomorphic to $\mathrm{End}({}_RV) = \mathrm{End}({}_{R/J(R)}V)$ by Proposition 1.11, we see from the discussion preceding the lemma that the ${}_{R/J(R)}V_{S/J(S)}$-dual takes simples to simples. But if ${}_RT$ or T_S is simple then $\mathrm{Hom}_R(T,E) \cong \mathrm{Hom}_{R/J(R)}(T,V)$ or $\mathrm{Hom}_S(T,E) \cong \mathrm{Hom}_{S/J(S)}(T,V)$ so the ${}_RE_S$-dual takes simples to simples.

Let ${}_RM$, N_S and ${}_RE_S$ are modules. A function

$$f : M \times N \longrightarrow E$$

is <u>a bilinear map</u> in case for all $m,m' \in M$, $n,n' \in N$, $r \in R$, and $s \in S$

$$\begin{aligned} f(m+m',n) &= f(m,n) + f(m',n) \\ f(m,n+n') &= f(m,n) + f(m,n') \\ f(rm,ns) &= rf(m,n)s. \end{aligned}$$

Now let $f : M \times N \longrightarrow E$ be a bilinear map. For each $A \subseteq M$ and $B \subseteq N$, the <u>right annilator of A in N</u> and the <u>left annilator of B in M</u> is defined to be

$$Ran_N(A) = \{n \in N \mid f(a,n) = 0 \text{ for all } a \in A\}$$

and

$$Lan_M(B) = \{m \in M \mid f(m,b) = 0 \text{ for all } b \in B\}.$$

If we let $M^* = Hom_R(M,E)_S$ and $N^* = {}_R Hom_S(N,E)$, we have two bilinear maps

$$M \times M^* \longrightarrow E \quad \text{and} \quad N^* \times N \longrightarrow E$$

defined by

$$(m,g) \longmapsto g(m) \quad \text{and} \quad (h,n) \longmapsto h(n),$$

respectively. Thus, if $A \subseteq M$ and $B \subseteq M^*$,

$$Ran_{M^*}(A) = \{g \in M^* \mid A \subseteq Ker(g) \}$$

$$Lan_M(B) = \cap \{Ker(g) \mid g \in B\}.$$

Thus it is clear that the annihilators with respect to the above bilinear map are exactly the same as defined in Section 2.

THEOREM 11.9. Let ${}_R E_S$ be a bimodule such that the ${}_R E_S$-dual takes simples to simples. Let

$$f : {}_R M \times N_S \longrightarrow {}_R E_S$$

be a bilinear map with $Lan_M(N) = 0$ and $Ran_N(M) = 0$. If either ${}_R M$ or N_S has finite length then

(1) For each $K \leq M$ and each $L \leq N$,

$$Lan_M(Ran_N(K)) = K \quad \text{and} \quad Ran_N(Lan_M(L)) = L;$$

(2) The induced mappings

$$g : {}_R M \longrightarrow N^* \quad \text{and} \quad h : N_S \longrightarrow M^*$$

defined by $g(m) : n \longmapsto f(m,n)$ and $h(n) : m \longmapsto f(m,n)$ are isomorphisms;

(3) All submodules and factor modules of M and N are E-reflexive;

(4) ${}_RE$ is M-injective and E_S is N-injective.

PROOF. Let W be a left R- or a right S-module of finite length and K a maximal submodule of W. The exact sequence

$$0 \longrightarrow K \longrightarrow W \longrightarrow W/K \longrightarrow 0$$

induces an exact sequence

$$0 \longrightarrow (W/K)^* \longrightarrow W^* \longrightarrow K^*$$

where $(W/K)^*$ is simple by the assumption. Hence we have $c(W^*) \leq c(K^*) + 1$. Using induction on the length of W, we see that

$$c(W^*) \leq c(W).$$

Moreover, the inequalities

$$c(W) \geq c(W^*) \geq c(W^{**})$$

imply that W is E-reflexive if and only if W is E-torsionless.

Now we suppose it is ${}_RM$ that has a composition series. For each $K \leq M$ define

$$p^K : N/\mathrm{Ran}_N(K) \longrightarrow K^*$$

via

$$p^K(n + \mathrm{Ran}_N(K)) : k \longmapsto f(k,n), \quad k \in K \text{ and } n \in N;$$

and for each $L \leq N$ define

$$q_L : \mathrm{Lan}_M(L) \longrightarrow (N/L)^*$$

via

$$q_L(m) : n + L \longmapsto f(m,n), \quad m \in \mathrm{Lan}_M(L) \text{ and } n \in N.$$

It is easy to check that both p^K and q_L are monomorphisms. These monomorphisms and the above discussions yield the following inequalities for each $K \leq M$:

$$\begin{aligned} c(K) &\leq c(\mathrm{Lan}_M\mathrm{Ran}_N(K)) \leq c((N/\mathrm{Ran}_N(K))^*) \\ &\leq c(N/\mathrm{Ran}_N(K)) \leq c(K^*) \leq c(K). \end{aligned}$$

These must all be equalities. Since they are, we see that

$$\mathrm{Lan}_M\mathrm{Ran}_N(K) = K$$

and that p^K is an isomorphism for all $K \leq M$. Since p^M : $N \cong M^*$, it follows by symmetry that (1) and (2) hold, and that M and N (being isomorphic to E-duals of one another), and all of their submodules, are E-torsionless (see Theorem 1.20 and the discussion preceding it). But also factor modules of N and, by symmetry, of M are also E-torsionless because of the isomorphisms $p^{\mathrm{Lan}_M(L)}$ which yield

$$N/L = N/\mathrm{Ran}_N\mathrm{Lan}_M(L) \cong \mathrm{Lan}_M(L)^*.$$

Thus since E-torsionless modules of fiite length are E-reflexive, (3) also holds. Finally, since p^K is an isomorphism, if $K \leq M$ every homomorphism $h : {}_RK \longrightarrow {}_RE$ is of the form

$$h : k \longmapsto f(k, n_h) \quad (k \in K)$$

for some $n_h \in N$; and such a homomorphism can be extended to

$$\bar{h} : m \longmapsto f(m, n_h) \quad (m \in M).$$

Thus ${}_RE$ is M-injective, by symmetry E_S is N-injective, and the proof is complete.

Suppose that eR_R and ${}_RE$ form a pair, and let $S = \mathrm{End}({}_RE)$. Then by Lemma 11.4 the bilnear map

$$f : {}_{eRe}eR \times E_S \longrightarrow {}_{eRe}eE_S$$

via multiplication

$$f : (er,x) \longmapsto erx$$

has the property that $\mathrm{Lan}_{eR}(E) = 0$ and $\mathrm{Ran}_E(eR) = 0$. Moreover by Lemma 11.6 ${}_{eRe}eE$ is a finitely cogenerated injective cogenerator, and by Lemma 11.5 S is canonically isomorphic to its endomorphism ring; so that by Lemma 11.8 the ${}_{eRe}E_S$-dual takes simples to simples. Thus we apply the above theorem to obtain

LEMMA 11.10. Let eR_R and ${}_RE$ form a pair over a

semiprimary ring R and let $S = \mathrm{End}({}_RE)$. If either ${}_{eRe}eR$ or E_S has finite length then

(1) For each left ideal $I \leq {}_RR$ and each right submodule $W \leq E_S$

$$\mathrm{Lan}_{eR}(\mathrm{Ran}_E(eI)) = eI \quad \text{and} \quad \mathrm{Ran}_E(\mathrm{Lan}_{eR}(W)) = W;$$

(2) The mappings

$$g : {}_RE_S \longrightarrow \mathrm{Hom}_{eRe}(eR,eE) \quad \text{and} \quad h : {}_{eRe}eR_R \longrightarrow \mathrm{Hom}_S(E,eE)$$

defined by the right and left multiplication

$$g(x) : er \longmapsto erx \quad \text{and} \quad h(er) : x \longmapsto erx$$

are bimodule isomorphisms.

The following theorem is due to Azumaya [59] and Morita [58] which gives several necessary and sufficient conditions on a left artinian ring R, a ring S (will be right artinian), and ${}_RE_S$ to insure a duality between the category of finitely generated left R-modules and that of finitely generated right S-modules.

THEOREM 11.11 [Azumaya, Morita]. Let R be a left artinian ring and let ${}_RE_S$ be a bimodule. Then the following statements are equivalent:

(1) The ${}_RE_S$-dual $(\ )^*$ defines a duality between the category of finitely generated left R-modules and that of finitely generated right S-modules;

(2) R, E and S satisfy

(i) S is right artinian,

(ii) All finitely generated left R-modules and right S-modules are E-reflexive;

(3) R, E and S satisfy

(i) S is right artinian,

(ii) ${}_RE$ and E_S are faithful,

(iii) All simple left R-modules and right S-modules are E-reflexive;

(4) R, E and S satisfy

(i) ${}_RE$ is finitely generated (S is right artinian),
(ii) ${}_RE$ and E_S are faithful,
(iii) The ${}_RE_S$-dual takes simples to simples;

(5) R, E and S satisfy
(i) S_S is E-reflexive,
(ii) ${}_RE$ is a finitely generated injective cogenerator;

(6) R, E and S satisfy
(i) S_S is E-reflexive,
(ii) ${}_RE$ and E_S are injective cogenerators;

(7) R, E and S satisfy
(i) ${}_RE$ is finitely generated (S is right artinian),
(ii) for each $I \leq {}_RR$ and each $V \leq E_S$,
$$\mathrm{Lan}_R(\mathrm{Ran}_E(I)) = I \quad \text{and} \quad \mathrm{Ran}_E(\mathrm{Lan}_R(V)) = V,$$
(iii) for each $K \leq S_S$ and each $W \leq {}_RE$,
$$\mathrm{Ran}_S(\mathrm{Lan}_E(K)) = K \quad \text{and} \quad \mathrm{Lan}_E(\mathrm{Ran}_S(W)) = W.$$

PROOF. (1) ==> (2). by Theorem 2.8.

(2) ==> (3). Since both ${}_RR$ and S_S are E-torsionless, both ${}_RE$ and E_S are faithful by Theorem 1.20.

(3) ==> (4). Let T be a simple left R-module. Then $T^{**} \neq 0$ implies $T^* \neq 0$. Since S is right artinian, T^* contains a maximal submodule M. The natural exact sequence
$$0 \longrightarrow M \xrightarrow{i_M} T^* \longrightarrow T^*/M \xrightarrow{n_M} 0$$
induces an exact sequence
$$0 \longrightarrow (T^*/M)^* \xrightarrow{n_M^*} T^{**} \xrightarrow{i_M^*} M^*.$$
Being the E-dual of a simple module, $(T^*/M)^* \neq 0$. Thus, since T^{**} is simple, n_M^* is an epimorphism and consequently $i_M^* = 0$. But T^* is E-torsionless (see Theorem 1.20) or, equivalently, there is an embedding $T^* \longrightarrow E^A$ for some set A. If $M \neq 0$, the embeddings $M \xrightarrow{i_M} T^* \longrightarrow E^A$ gives a non-zero homomorphism

$M \xrightarrow{i_M} T^* \longrightarrow E$, which contradicts to that $i_M^* = 0$. Hence $M = 0$. We prove that T^* is simple for every simple left R- (and, similarly, every simple right S-) module T. Now we see that (3) implies both the parenthetical and the non-parenthetical version of (4) by applying Theorem 11.9 to the bilinear map

$$_RE \times S_S \longrightarrow {}_RE_S$$

given by the scalar multiplication $(x,s) \longmapsto xs$.

(4) ==> (5). This implication follows from Theorem 11.9 applied to the bilinear maps

$$_RE \times S_S \longrightarrow {}_RE_S \quad \text{and} \quad {}_RR \times E_S \longrightarrow {}_RE_S$$

given by scalar multiplication. The first application shows that the right multiplication $S \longrightarrow \mathrm{End}(_RE)$ is an isomorphism, hence S_S is E-reflexive. The second shows that $_RE$ is $_RR$-injective, hence $_RE$ is injective. Since each simple left R-module is E-reflexive (Theorem 11.9(3)), hence E-torsionless. Then $_RE$ contains each simple left R-module, so $_RE$ is a cogenerator (Proposition 1.13).

(5) ==> (6). By Lemma 11.8, Theorem 11.9 applied to scalar multiplication

$$_RE \times S_S \longrightarrow {}_RE_S$$

gives the proof this implication.

(6) ==> (1). Since S_S is E-reflexive, the right multiplication $S \longrightarrow \mathrm{End}(_RE)$ is an isomorphism. Then $_RE$ is E-reflexive. The injectivity of E_S implies that $_RE$ is linearly compact by Theorem 4.1. So $_RE$ is finitely generated (Corollary 3.5), since R is left artinian. Therefore $_RE_S$ defines a duality (Theorem 4.5).

(4) ==> (7). By Theorem 11.9.

(7) ==> (4). Since $\mathrm{Lan}_R(E) = \mathrm{Lan}_R\mathrm{Ran}_E(0) = 0$ and $\mathrm{Ran}_S(E) = \mathrm{Ran}_R\mathrm{Lan}_E(0) = 0$, both $_RE$ and E_S are faithful. If I is a maximal left ideal of R then $\mathrm{Ran}_E(I)$ is a minimal submodule of E_S. But it is easy to see that

$(R/I)^* \cong \mathrm{Ran}_{R^*}(I) \cong \mathrm{Ran}_E(I)$ for any left ideal I of R (see Theorem 1.21(2)). This and a symmetric argument show that the ${}_R E_S$-dual takes simple to simples.

Let R be a semiprimary ring with $J = J(R)$. If M is a left R-module and n is the smallest positive integer with $J^n M = 0$, the <u>upper Loewy series for M</u> is

$$M > JM > J^2M > \dots > J^nM = 0$$

and the <u>lower Loewy series for M</u> is

$$0 < \mathrm{Ran}_M(J) < \mathrm{Ran}_M(J^2) < \dots < \mathrm{Ran}_M(J^n) = M.$$

Letting $J^0 = R$, $J^{k-1}M/J^kM$ is the <u>k-th upper Loewy factor of M</u> and $\mathrm{Ran}_M(J^k)/\mathrm{Ran}_M(J^{k-1})$ is the <u>k-th lower Loewy factor of M</u>. Lowey series and factors for a right R-module are defined in the obvious manner. The next Fuller's result [69] gives a natural one-to-one correspondence between the homogeneous components of the k-th upper (lower) Loewy factor of E and the k-th lower (upper) Loewy factor of eR whenever eR and E form a pair over a left artinian ring R. We state the theorem in terms of indecomposable injectives and projectives.

THEOREM 11.12. Let E be an indecomposable injective left module over a left artinian ring R. Let e be a primitive idempotent of R with $T(Re) \cong \mathrm{Soc}(E)$. If e' is any primitive idempotent of R then $T(Re')$ appears in the k-th upper (lower) Loewy factor of E if and only if $T(e'R)$ appears in the k-th lower (upper) Loewy factor of eR.

PROOF. First we show that if $I \subseteq K$ are ideals of R then

$$\mathrm{Lan}_R(\mathrm{Ran}_E(I)/\mathrm{Ran}_E(K)) = \mathrm{Ran}_R(eK/eI).$$

It follows from Lemma 11.5(2) that

$$\mathrm{Ran}_R(eK/eI) \subseteq \mathrm{Lan}_R(\mathrm{Ran}_E(I)/\mathrm{Ran}_E(K)).$$

On the reverse inclusion, suppose that eKr is not contained in eI for some $r \in R$. Then $(eKr + eI)/eI$ has a simple eRe-factor and using Lemma 11.6 we see that there is a map $f \in \mathrm{Hom}_{eRe}(eK/eI, eE)$ such that $(rf)(eK/eI) = f((eKr + eI)/eI) \neq 0$. Thus by Lemma 11.5(2), r does not annihilate $\mathrm{Ran}_E(I)/\mathrm{Ran}_E(K)$ and the desired equality holds.

Now let $I = J^k \subseteq J^{K-1} = K$ we see that if e' is a primitive idempotent of R then $T(Re')$ can be embedded in $\mathrm{Ran}_E(J^k)/\mathrm{Ran}_E(J^{k-1})$ if and only if $T(e'R)$ can be embedded in eJ^{k-1}/eJ^k.

On the other hand, by Lemmas 11.10 and 11.4(1),

$$\begin{aligned} J^kE &= \mathrm{Ran}_E(\mathrm{Lan}_{eR}(J^kE) = \mathrm{Ran}_E(\mathrm{Lan}_{eR}(J^k) \\ &= \mathrm{Ran}_E(e\mathrm{Lan}_R(J^k) = \mathrm{Ran}_E(\mathrm{Lan}_R(J^k). \end{aligned}$$

So that, letting $I = \mathrm{Lan}_R(J^{k-1}) \subseteq \mathrm{Lan}_R(J^K) = K$,

$$\begin{aligned} \mathrm{Lan}_R(J^{k-1}E/J^kE) &= \mathrm{Lan}_R(\mathrm{Ran}_E(\mathrm{Lan}_R(J^{k-1}))/\mathrm{Ran}_E(\mathrm{Lan}_R(J^k))) \\ &= \mathrm{Ran}_R(e\mathrm{Lan}_R(J^k)/e\mathrm{Lan}_R(J^{k-1})) \end{aligned}$$

and the theorem is proved.

THEOREM 11.13 [Fuller]. Let e and f be idempotents in a left or right artinian ring R. Then the following are equivalent:

(1) Rf is injective and eR and $Rf = E$ form a pair;

(2) For each f_i in a basic set of idempotents for fRf there is a primitive idempotent e_i in R such that

$$\mathrm{Soc}(Rf_i) \cong T(Re_i) \quad \text{and} \quad \mathrm{Soc}(e_iR) \cong T(f_iR);$$

(3) Rf and eR satisfy

(i) $\mathrm{Ran}_{Rf}(eR) = 0 = \mathrm{Lan}_{eR}(Rf)$,

(ii) The bimodule ${}_{eRe}eRf_{fRf}$ defines a duality between the category of finitely generated left eRe-modules and the category of finitely generated right fRf-modules;

(4) eR is injective and Rf and $eR = E'$ form a pair.

PROOF. (1) ==> (2). In this case we may assume that e is primitive and that $Rf \cong E(T(Re))$. Since R is left or right artinian and $fRf \cong End({}_RRf)$ canonically, it follows from Lemma 11.10 that both ${}_{eRe}eR$ and Rf_{fRf} have finite length and that

$${}_{eRe}eR_R \cong Hom_{fRf}(Rf, eRf).$$

Then eRe is left artinian and ${}_{eRe}eRf$ is finitely generated. By Lemma 11.6, ${}_{eRe}eRf$ is an injective cogenerator. We know that for any left R-module ${}_RM$, the bimodule ${}_{BiEnd({}_RM)}M_{End({}_RM)}$ is a faithful balanced bimodule, where $BiEnd({}_RM) = End(M_{End({}_RM)})$. Hence it follows from Lemma 11.5 that the multiplication $fRf \longrightarrow End({}_{eRe}eRf)$ is an isomorphism. From Theorem 11.11(5) we see that ${}_{eRe}eRf_{fRf}$ induces a duality. In particular, $eRf_{fRf} \cong ({}_{eRe}eRe)^* \cong E(T(fRf_{fRf}))$. Applying the right-left symmetric versions of Lemmas 11.6 and 11.10(2) we have

$$eRf \cong E(T(fR))f$$

as right fRf-modules and

$$E(T(fR) \cong Hom_{fRf}(Rf, eRf)$$

over R. Combining this with the first isomorphism of the proof we see that

$$E(T(fR)) \cong eR.$$

(2) ==> (3). If f_i and e_i, $i=1,\ldots,n$, satisfy the condition (2) we may assume $e_1,\ldots,e_n$ are also pairwise orthgonal. Let $e = e_1+\ldots+e_n$. Then no minimal right ideal in eR anihilates f and no minimal left ideal in Rf annihilates e, so (i) holds. This insures that eRf_{fRf} and ${}_{eRe}eRf$ are faithful. Since R is either left or right artinian, both ${}_{eRe}eRe$ and ${}_{eRe}eRf$ or both fRf_{fRf} and eRf_{fRf} have finite length. Using Theorem 11.11(4) we only need to show that the ${}_{eRe}eRf_{fRf}$-dual takes simples to simples. To do so, let e' be a primitive idempotent in eRe, so $T(eRe')$ is a typical simple

left eRe-module. Then we have fRf-isomorphisms

$$\begin{aligned}\mathrm{Hom}_{eRe}(T(eRe'), eRf) &\cong \mathrm{Hom}_{eRe}(T(eRe'), \mathrm{Soc}({}_{eRe}eRf)) \\ &\cong \mathrm{Hom}_{eRe}(eRe', \mathrm{Soc}({}_{eRe}eRf)) \\ &\cong e'\cdot \mathrm{Soc}({}_{eRe}eRf) \\ &= e'\mathrm{Lan}_R(J)f\end{aligned}$$

where $J = J(R)$, and the last equality follows from the next result.

Sublemma. Under the assumption of (2), it holds that $\mathrm{Soc}({}_{eRe}eRf) = e\mathrm{Ran}_R(J)f = e\mathrm{Lan}_R(J)f = \mathrm{Soc}(eRf_{fRf})$.

Proof. Let $W = eWf = \mathrm{Soc}({}_{eRe}eRf)$. Then, since eRe/eJe is semisimple and $eJe\mathrm{Ran}_R(J)f \subseteq J\mathrm{Ran}_R(J)f = 0$, we know that $e\mathrm{Ran}_R(J)f \subseteq W$. But also $eRJW = eJeWf = eJeW = 0$ so $JW \subseteq \mathrm{Ran}_{Rf}(eR) = 0$. So we have the first equality $W = e\mathrm{Ran}_R(J)f$. Now, by symmetry, the proof will be complete once we show that $e\mathrm{Ran}_R(J)f \subseteq e\mathrm{Lan}_R(J)f$. If not, i.e., suppose that $e\mathrm{Ran}_R(J)fJ \neq 0$. Then there exist primitive idempotents f' and e' in fRf and eRe, respectively, such that

$$e'\mathrm{Ran}_R(J)f'J \neq 0.$$

Because $0 \neq e'\mathrm{Ran}_R(J)f' = e'\cdot\mathrm{Soc}({}_RRf')$, we have

$$\mathrm{Soc}({}_RRf') \cong Re'/Je' \quad\text{and}\quad \mathrm{Soc}(e'R_R) \cong f'R/f'J.$$

Then the right ideal $e'\mathrm{Ran}_R(J)f'J \leq e'R$ must contain a copy of $f'R/f'J$; so that $\mathrm{Ran}_R(J)f'Jf' \neq 0$, and the right multiplication by some $jf' \in Jf'$ induces a monomorphism

$$0 \longrightarrow Rf' \longrightarrow Jf'.$$

This contradicts the fact that Rf' have finite Loewy length, and the Sublemma is proved.

Now we complete the proof of (3)(ii): Since $e'\mathrm{Lan}_R(J) = \mathrm{Lan}_{e'R}(J) = \mathrm{Soc}(e'R_R)$ is simple over R and is not annihilated by f this module must be simple over fRf. A symmetric argument completes the proof.

(3) ==> (1). Let Rf and eR satisfy the conditions of (3), and write $(\ )^* = \mathrm{Hom}(-, eRf)$. Using the condition (i) we see that

right and left multiplications define bimodule monomorphisms

$$g: {}_R Rf_{fRf} \longrightarrow (eR)^* \quad \text{and} \quad h: {}_{eRe}eR_R \longrightarrow (Rf)^*.$$

Then, since one of ${}_{eRe}eR$ and Rf_{fRf} has finite length, both are eRf-reflexive and have finite length (Theorem 2.8); and since a Morita duality preserves composition length (Theorem 2.6), both g and h must be isomorphisms. Since it is injective cogenerator, we can write

$${}_{eRe}eRf \cong \oplus\hat{E}_{ij} \quad \text{with} \quad \hat{E}_{ij} \cong E(T(eRe_i) \quad (i = 1,\ldots,m)$$

where $e_1,\ldots,e_m$ is a basic set of idempotents for eRe. Thus, by Lemma 11.6, if

$${}_RE = \oplus E_{ij} \quad \text{with} \quad E_{ij} \cong E(T(Re_i) \quad (i = 1,\ldots,m)$$

then eR_R and ${}_RE$ form a pair and

$$eRf \cong eE$$

over eRe. Now applying Lemma 11.10(2) we have

$$Rf \cong \mathrm{Hom}_{eRe}(eR, eRf) \cong E;$$

and the first three conditions are equivalent.

The last three conditions are equivalent by symmetry.

Next we give several characterizations of QF-rings as applications of Fuller's Theorem. An artinian cogenerator ring R is called a <u>quasi-Frobenius ring</u> (or QF-ring), i.e., the ${}_RR_R$-dual defines a duality between the category of finitely generated left and right R-modules. QF-rings were introduced by Nakayama [39] and the condition (3) in Theorem 11.15 is his original definition.

LEMMA 11.14. If R satisfies the double annihilator property, then the following statements are equivalent:

(1) R is left noetherian;

(2) R is left artinian;

(3) R is right noetherian;

(4) R is right artinian.

PROOF. Since left (right) artinian rings are left (right) noetherian, by symmetry we only need to verify the implication (1) ==> (4). Now $I = Lan_R Ran_R(I)$ and $K = Ran_R Lan_R(K)$ for each left ideal I and right ideal K of R, hence $I \longmapsto Ran_R(I)$ and $K \longmapsto Lan_R(K)$ are inverse lattice antiisomorphisms between the left ideals and right ideals. So it is easy to see that (1) implies (4).

For more details about QF-rings, the reader is refered to Faith [76, Chapter 24].

THEOREM 11.15. Let R be a left or right artinian ring. The following statements are equivalent:

(1) R is a QF-ring;

(2) ${}_RR$ is injective;

(3) For each primitive idempotent f in R there is a primitive idempotent e in R such that

$$Soc(Rf) \cong T(Re) \quad \text{and} \quad Soc(eR) \cong T(fR);$$

(4) R satisfies the double annihilator property;

(5) The ${}_RR_R$-dual defines a duality between the category of finitely generated left and right R-modules;

(6) The ${}_RR_R$-dual $(\ )^*$ takes simples to simples.

PROOF. (1) <==> (2) <==> (3) <==> (5). Theorem 11.13.

(5) <==> (6). Theorem 11.11.

(1) ==> (4). Theorem 5.4.

(4) ==> (2). Lemma 11.14, Propositions 5.2 and 5.3.

In view of the condition (6) in the above theorem, we present the following Dieudonné's Theorem [58] that we shall use in the next chapter.

THEOREM 11.16 (Dieudonné). If R is an artinian ring with

$c({}_RR) = c(R_R)$, and for each simple left R-module ${}_RT$ its dual $T^* = \mathrm{Hom}_R(T,R)$ has legth ≤ 1, then R is a QF-ring.

Before we prove the theorem, we need two facts and a lemma. If ${}_RN$ is a submodule of ${}_RM$, we have an exact sequence of right R-modules

$$0 \longrightarrow (M/N)^* \longrightarrow M^* \longrightarrow N^*.$$

where $(-)^*_R = \mathrm{Hom}_R(-,{}_RR)$. Thus we have a natural isomorphism,

(i) $$(M/N)^* \cong \mathrm{Lan}_{M^*}(N),$$

and a natural monomorphism

(ii) $$M^*/\mathrm{Lan}_{M^*}(N) \longrightarrow N^*,$$

LEMMA 11.17. Under the assumption of Theorem 11.16, $c(M^*_R) \leq c({}_RM)$ for any left R-module ${}_RM$ of finite length.

PROOF. The proof is by induction on $n = c({}_RM)$, the result being true by assumption for $n = 1$. Let N be a non-zero and proper submodule of M, then $c(N) < n$ and $c(M/N) < n$. Hence

$$c(\mathrm{Lan}_{M^*}(N)) \overset{(i)}{=} c(M/N)^* \leq c(M/N)$$

and

$$c(M^*/\mathrm{Lan}_{M^*}(N)) \overset{(ii)}{\leq} c(N^*) \leq c(N).$$

It follows that

$$c(M^*_R) = c(\mathrm{Lan}_{M^*}(N)) + c(M^*/\mathrm{Lan}_{M^*}(N)) \leq c(M/N) + c(N) = c({}_RM).$$

PROOF of THEOREM 11.16. Using Theorem 11.15, we show that ${}_RR$ is injective. Let I be a left ideal of R. We shall prove that the exact sequence of left R-modules

$$0 \longrightarrow I \overset{i}{\hookrightarrow} R \longrightarrow R/I \longrightarrow 0$$

induces an exact sequence of ringt R-modules

$$0 \longrightarrow (R/I)^* \longrightarrow R^* = R_R \xrightarrow{i^*} I^* \longrightarrow 0.$$

So we need to show that i^* is an epimorphism, i.e., the natural monomorphism

(ii) $$R/\mathrm{Ran}_R(I) \longrightarrow I^*$$

is an isomorphism. By (ii) and Lemma 11.17, we have $c(R/\mathrm{Ran}_R(I)) \leq c(I^*) \leq c(I)$. In the above situation, we have

(i) $$(R/I)^* \cong \mathrm{Ran}_R(I).$$

Hence

$$\begin{aligned} c(R_R) &= c(R/\mathrm{Ran}_R(I)) + c(\mathrm{Ran}_R(I)) \\ &\leq c(I) + c(R/I)^* \\ &\leq c(I) + c(R/I) \qquad \text{(Lemma 11.17)} \\ &= c({}_RR) = c(R_R). \end{aligned}$$

It follows that $c(I) = c(R/\mathrm{Ran}_R(I)) \leq c(I^*) \leq c(I)$, and hence $c(R/\mathrm{Ran}_R(I)) = c(I^*)$. Therefore the monomorphism (ii) is an isomorphism.

We conclude this section by raising the following

QUESTION 11.18. If R is an artinian ring with a duality induced by ${}_RE_S$ then S is a right artinian ring. Is S also left artinian?

SECTION 12. ARTINIAN LEFT DUO RINGS

A ring R is called <u>left</u> (<u>right</u>) <u>duo</u> in case each left (right) ideal of R is a two-sided ideal, and it is called <u>duo</u> if it is both left and right duo. It is easy to see that R is left (right) duo if and only if $rR \subseteq Rr$ $(Rr \subseteq rR)$ for each $r \in R$. The following example was given in Rosenberg and Zelinsky [59].

EXAMPLE 12.1. Let K be a field with a monomorphism f into itself such that $\dim({}_{f(K)}K) = \infty$; e.g., $K = F(X_1, X_2, \ldots)$ with F a field, $f(X_i) = X_{i+1}$ and $f =$ identity on F. Let ${}_KM = {}_KK$ as left K-module, and as right K-module: $m \cdot k = f(k)m$ in K, where $m \in M$ and $k \in K$. Then it is easy to see that $R = K \propto M$ is a local left artinian left duo ring such that $R/J(R)$ is a field. We show that ${}_K\mathrm{Hom}_K({}_KM_K, {}_KK)$ is not finitely generated. In fact for any finitely many $g_1, \ldots, g_n \in {}_K\mathrm{Hom}_K({}_KM_K, {}_KK)$, let $g_i(1) = k_i$, $i = 1, \ldots, n$. Since M_K is not finitely generated, there is an $k \in M$ such that $k \notin \Sigma_{i=1}^n k_iK$. Define $g : {}_KM \longrightarrow {}_KK$ via $m \longmapsto mk$, then $g \notin \Sigma_{i=1}^n Kg_i$. Hence ${}_K\mathrm{Hom}_K({}_KM_K, {}_KK)$ is not finitely generated and $R = K \propto M$ does not have a duality by Theorem 10.1.

In the above example, R is a local left artinian left duo ring such that $R/J(R)$ is a field. We note that R is not right artinian. Motivated by this example we study (two-sided) artinian left duo rings in this section and show that there is a large class of artinian left duo rings that possess dualities.

LEMMA 12.2. If e is an idempotent element of a left duo ring R, then e belongs to the center of R.

PROOF. Since $eR \subseteq Re$, $eR(1-e) \subseteq Re(1-e) = 0$. Hence $er = ere$ for each $r \in R$. Similarly, $re - ere \in (1-e)Re \subseteq R(1-e)e = 0$. So $er = ere = re$ for each $r \in R$.

Using the characterizations of semiperfect rings in Theorem 1.18, we see that a semiperfect left duo ring is a finite direct sum of local left duo rings. Therefore an artinian left duo ring is a finite direct sum of local artinian left duo rings. Hence we consider a local artinian left duo ring R with radical $J \neq 0$.

According ot Corollary 11.3, we may assume that $J^2 = 0$. Let $D = R/J$ be the division ring. Then ${}_DJ_D$ is a bi-vector space with finite dimensional on each side, since R is artinian. By Corollary 11.3, R has a duality if and only if ${}_D\mathrm{Hom}_D(J,D)$ is finitely generated.

Since R is left duo, $aD \subseteq Da$ for each $0 \neq a \in J$, so there is a ring monomorphism

$$\sigma_a : D \longrightarrow D$$

given by $\sigma_a(d)a = ad$ for all $d \in D$. And $\sigma_a(D)$ is a division subring of D.

LEMMA 12.3. For each $0 \neq a \in J$, $D_{\sigma_a(D)}$ is finite dimensional.

PROOF. Since R is artinian left duo, $(Da)_D$ is finitely generated right D-module. Let $Da = \oplus_{j=1}^m (d_ja)D = \oplus_{j=1}^m d_j\sigma_a(D)a$ for some $d_1,\ldots,d_m \in D$, and then $D = \Sigma_{j=1}^m d_j\sigma_a(D)$.

Xue [89c] proved the following result that gives a complete characterization of artinian left duo rings via Morita duality.

THEOREM 12.4. For the local artinian left duo ring R with $J^2 = 0$ we have the following:

(1) If $\dim({}_DJ) > 1$, then R has a duality; in this case D is a field;

(2) If $\dim({}_DJ) = 1$; let $J = Da$ and $\sigma = \sigma_a$; then

(i) if ${}_{\sigma(D)}D$ is finite dimensional, R has a duality; in particular, if D is a field then R has a duality;

(ii) if ${}_{\sigma(D)}D$ is not finite dimensional, R does not have a duality.

PROOF. (1) First we use an idea of Habeb [89] to show that the monomorphism σ_a does not depend on a.

Let a and $b \in J$ be linearly independent over D. Then for each $d \in D$, we have

$$\sigma_a(d)a = ad, \quad \sigma_b(d)b = bd, \quad \text{and} \quad \sigma_{a+b}(d)(a+b) = (a+b)d.$$

So $\sigma_a(d) = \sigma_{a+b}(d) = \sigma_b(d)$, and hence $\sigma_a = \sigma_b$.

Now let $0 \neq a$ and $0 \neq b \in J$ be linearly dependent over D. Since $\dim({}_DJ) > 1$, we can find $c \in J$ such that c is linearly independent of a whence linearly independent of b. From the discussion above, we have $\sigma_a = \sigma_c = \sigma_b$.

This proves that all the non-zero elements determine the same monomorphism, say $\sigma : D \longrightarrow D$. Let $0 \neq d_1, d_2 \in D$ and $0 \neq a \in J$, then by the uniqueness of σ we have

$$d_1\sigma(d_2)a = d_1ad_2 = \sigma(d_2)d_1a.$$

So $d_1\sigma(d_2) = \sigma(d_2)d_1$, which says that $\sigma(D)$ is contained in the center of D. In particular, $\sigma(D)$ is a field. But $D \cong \sigma(D)$, so D is a field. It follows from Lemma 12.3 that ${}_{\sigma(D)}D$ is finitely generated.

Let $D = \Sigma_{j=1}^m \sigma(D)d_j$ and $J = \oplus_{i=1}^n Da_i$ for some $d_1,\ldots,d_m \in D$ and $a_1,\ldots,a_n \in J$. By the uniqueness of σ, we have $\sigma(d)a_i = a_id$ for all $d \in D$ and all i, $1 \leq i \leq n$. Since $a_iD \subseteq Da_i$, we have

$${}_D\mathrm{Hom}_D(J,D) \cong \oplus_{i=1}^n \mathrm{Hom}_D(Da_i\ ,D)$$

as left D-modules. So we only need to show that each ${}_D\mathrm{Hom}_D(Da_i,D)$ is finitely generated. To do so, let $a = a_i$ be fixed.

We define $f_j \in \mathrm{Hom}_D(Da,D)$ via $f_j(a) = d_j$, $1 \leq j \leq m$, and claim that ${}_D\mathrm{Hom}_D(Da,D) = \Sigma_{j=1}^m Df_j$. In fact, for each $f \in \mathrm{Hom}_D(Da,D)$, we have $f(a) = \Sigma_{j=1}^m \sigma(t_j)d_j$ for some $t_j \in D$. Then

$$(\Sigma_{j=1}^m t_jf_j)(a) = \Sigma_{j=1}^m f_j(at_j) = \Sigma_{j=1}^m f_j(\sigma(t_j))(a)$$
$$= \Sigma_{j=1}^m \sigma(t_j)f_j(a) = \Sigma_{j=1}^m \sigma(t_j)d_j = f(a).$$

Therefore $f = \Sigma_{j=1}^{m} t_j f_j \in \Sigma_{j=1}^{m} Df_j$, which proves our claim. It follows that ${}_D\mathrm{Hom}_D(J,D)$ is finitely generated.

(2)(i) Since ${}_{\sigma(D)}D$ is finitely generated, we let $D = \Sigma_{j=1}^{m} \sigma(D)d_j$ for some $d_1, \ldots, d_m \in D$. Defining the same f_j's as that given in the proof of (1), one shows that

$$ {}_D\mathrm{Hom}_D(J,D) = {}_D\mathrm{Hom}_D(Da,D) = \Sigma_{j=1}^{m} Df_j , $$

which is finitely generated. If D is a field, then by Lemma 12.3 ${}_{\sigma(D)}D$ is finite dimensional.

(2)(ii) For $f_1, \ldots, f_n \in \mathrm{Hom}_D(Da,D)$, let $f_i(a) = d_i$. Since ${}_{\sigma(D)}D$ is not finitely generated, there is a $d \in D\backslash(\Sigma_{i=1}^{m}\sigma(D)d_i)$. Define $f \in \mathrm{Hom}_D(Da,D)$ via $f(a) = d$. Then $f \notin \Sigma_{i=1}^{m} Df_i$ and so ${}_D\mathrm{Hom}_D(Da,D)$ is not finitely generated. It follows that R does not have a duality.

It should be noted that artinian left duo rings arise much more frequently than artinian (two-sided) duo rings. A typical example is Dlab and Ringel's "exceptional (1,2) ring [72, Proposition II.3.3]. These rings belong to the class of artinian left duo rings described in Theorem 12.4(2)(i). An easiest example is the following

EXAMPLE 12.5. Let F be a field and let $K = F(X)$ be the field of rational functions over F. Let ${}_KV = {}_KK$ as left K-space. As right K-space, we define $v \cdot f(X) = f(X^2)v$ in K, where $v \in V$ and $f(X) \in K$. Then ${}_KV_K$ is a bi-vector space with $\dim({}_KV) = 1$ and $\dim(V_K) = 2$. Now let $R = K \propto V$ be the trivial extension of K by V. Then R is a local artinian left duo ring that is not right duo.

Using Cohn's division ring constructions [66], one constructs an artinian ring $\begin{bmatrix} D & D \\ 0 & C \end{bmatrix}$ without a duality (see Remark 2.9). The author is unable to answer the following

Question: Does there exist a division ring D with a division subring C such that $\dim(D_C)$ is finite, $\dim({}_CD) = \infty$, and $D \cong C$ as rings?

If the answer is yes, using the same method as in Example 12.1 we can construct a local artinian left duo ring without a duality. Simply let ${}_DM = {}_DD$ as left D-module, and as right D-module, define $m \cdot d = f(d)m$ in D where $f : D \cong C$. Then $R = D \propto M$ is a ring we need. In case the answer is no, the set of the rings in Theorem 12.4(2)(ii) is empty. In this event we conclude that every artinian left duo ring has a duality. Sumarizing this we get the following interesting relation between Morita duality and division ring extensions.

THEOREM 12.6. The following two statements are equivalent:

(1) Every artinian left duo ring has a Morita duality;

(2) There does not exist division ring extension $D \geqq C$ such that $\dim(D_C)$ is finite, $\dim({}_CD) = \infty$, and $D \cong C$ as rings.

According to Theorem 2.8, a one-sided artinian ring can not have self-duality. Schofield [85a, 85b] has shown that there are division ring extensions $D \geqq C$ such that both $\dim(D_C)$ and $\dim({}_CD)$ are finite but different. It follows from this that

EXAMPLE 12.7. The artinian ring $R = \begin{bmatrix} D & D \\ 0 & C \end{bmatrix}$ does have a duality but does not have self-duality: By Corollary 10.5, R has a duality induced by ${}_RE = \begin{bmatrix} D & 0 \\ D & C \end{bmatrix}$. It is easy to see that ${}_RE$ is the minimal cogenerator in R-Mod. Now

$$c(R_R) = 2 + \dim(D_C) \neq 2 + \dim({}_CD) = c({}_RE),$$

hence $R \not\cong \mathrm{End}({}_RE)$. Since R is a basic ring, R can not have a self-duality.

Using the following profound results of Schofield [85b, p.214-217] and Dowbor, Ringel and Simson [80], we can construct a local artinian left duo ring with a duality but without a self-duality. Example 12.9 was given in Xue [89d].

THEOREM 12.8 (Schofield [85b, p.214-217] and Dowbor, Ringel and Simson [80]). There is a division ring extension $D \geq C$ such that $\dim(D_C) = 2$, $\dim({}_CD) = 3$, and $f : D \cong C$ as rings.

EXAMPLE 12.9. Let ${}_DM = {}_DD$ as left D-module, and as right D-module, define $m \cdot d = f(d)m$ in D where $f : D \cong C$. Then $R = D \propto M$ is a local artinian ring with $c({}_RR) = 2$ and $c(R_R) = 3$, so R is left duo but not right duo. Let ${}_RE$ be the indecomposable injective R-module, then by Lemma 2.10 we have

$$E/\mathrm{Soc}({}_RE) \cong {}_R\mathrm{Hom}_R({}_RJ(R)_R, \mathrm{Soc}({}_RE))$$

whose length = $\dim({}_D\mathrm{Hom}_D({}_DD_C, {}_DD)) = \dim({}_CD) = 3$. It follows that $c({}_RE) = 4 \neq c(R_R)$. Hence R has a duality but does not have a self-duality.

CHAPTER 4
ARTINIAN RINGS (II) - AZUMAYA'S EXACT RINGS

Azumaya [83] initiated the study of exact rings and proved that exact rings have dualities. In this chapter, we present updated results about exact rings and Azumaya's conjecture that exact rings have self-dualities. In the first two sections, some properties of exact rings are formulated; in particular, an exact ring has a duality (Azumaya [83]) and the endomorphism ring of a finitely cogenerated injective cogenerator over an exact ring is also an exact ring (Habeb [89]). In Section 15, we consider locally distributive rings that are exact by Camillo, Fuller and Haack [86]. Although the self-duality for this class rings is still open this time, partial results will be given. And the self-duality for serial rings has been verified by Dischinger and Müller [84], and Waschbüsch [86] independently. In Section 16, we study artinian duo rings (that are exact by Habeb [89]), and prove Habeb's conjecture that the endomorphism ring of the minimal left cogenerator over an artinian duo ring is still an artinian duo ring and that an artinian duo ring has self-duality if its Jacobson radical is a direct sum of colocal ideals (Xue [89]). The results in this chapter provide further evidence for Azumaya's conjecture.

SECTION 13. AZUMAYA'S EXACT RINGS

Azumaya [83] introduced the notations of exact rings, proved that exact rings have dualities, and conjectured that they have self-duality. This conjecture is far from being solved, but we do give some partial answers in this section and the rest of this chapter. All results in this section are due to Azumaya [83].

Throughout this section, let R be a left artinian ring with radical J and a composition series of (two-sided) ideals

$$R = I_0 > I_1 > \ldots > I_{s-1} > I_s = 0.$$

R is called an <u>exact ring</u> in case for each i, every endomorphism of the left R-module I_{i-1}/I_i is given by the right-multiplication of an element of R. The notion of the exactness is independent of the choice of the above composition series and is left-right symmetric; in particular every exact ring is right artinian too.

Let $\bar{R} = R/J$ and $\bar{R} = \bar{R}_1 \oplus \bar{R}_2 \oplus \ldots \oplus \bar{R}_\ell$ the direct decomposition of $\bar{R}$ into simple components. For each i let e_i be a primitive idempotent element of R such that $\bar{e}_i = e_i + J \in \bar{R}_i$. So every simple left R-module is isomorphic to some $\bar{R}\bar{e}_i$, and $\bar{R}\bar{e}_i \cong \bar{R}\bar{e}_j$ if and only if $i = j$. Now each $\bar{e}_i$ is also a primitive idempotent of $\bar{R}$, and the left R-module $\bar{R}_i$ is the direct sum of, say $n(i)$ copies of $\bar{R}\bar{e}_i$. Moreover, R is a direct sum of indecomposable left ideals each of which is isomorphic to one of Re_i's and the multiplicity of Re_i in the decomposition is $n(i)$, i.e., we have ${}_R R \cong \oplus\, {}_R(Re_i)^{n(i)}$. The similar facts are also true for simple right R-modules $\bar{e}_i\bar{R}$ and indecomposable right ideals e_iR; in particular, we have $(\bar{R}_i)_R \cong (\bar{e}_i\bar{R})_R^{n(i)}$ and $R_R \cong \oplus\, (e_iR)_R^{n(i)}$.

Let M be a simple two sided R-module. Since J is nilpotent, M is then annihilated by J on both left- and right-hand sides and so can be regarded as a simple two sided $\bar{R}$-module. Since furthermore $M = \bar{R}M = \Sigma_i \bar{R}_iM$ and each $\bar{R}_iM$ is a two-sided $\bar{R}$-submodule of M, it follows that $\bar{R}_kM \neq 0$ whence $\bar{R}_kM = M$ for some k: but then $\bar{R}_iM = \bar{R}_i\bar{R}_kM = 0$ for every $i \neq k$. Thus k is the only index such that $\bar{R}_kM = M$. Similarly, there is a unique index t such that $M\bar{R}_t = M$, and we have $M\bar{R}_i = 0$ whenever $i \neq t$. This means that M can actually be regarded as a simple two sided $\bar{R}_k$-$\bar{R}_t$-module. We shall call $\bar{R}_k$ and $\bar{R}_t$ the <u>left</u> and the <u>right simple components</u> belonging to M, respectively.

Since $\bar{R}_t$ is a simple ring, $\bar{R}_t$ is considered as a subring of the endomorphism ring D of the left $\bar{R}_k$-module M. We call M <u>exact</u> if $\bar{R}_t = D$, i.e., if every endomorphism of the left $\bar{R}_k$-module M is given by the right-multiplication of an element of R. Now since $\bar{R}_k$ is a direct sum of simple left ideals isomorphic to $\bar{R}\bar{e}_k$, the left $\bar{R}_k$-module M is also a direct sum of copies of $\bar{R}\bar{e}_k$, which implies clearly that M is a progenerator. Therefore, it follows from Corollary 1.16 that M is a progenerator with respect to both $\bar{R}_k$ and $\bar{R}_t$ and also M is a balanced bimodule. Thus the notion of the exactness for M is left-right symmetric.

LEMMA 13.1. Let M be an exact simple two-sided R-module with left simple component $\bar{R}_k$ and right simple component $\bar{R}_t$. Then

(i) ${}_RMe_t \cong {}_R\bar{R}\bar{e}_k$, $e_kM \cong \bar{e}_t\bar{R}_R$,

(ii) ${}_R\mathrm{Hom}({}_RM, {}_R\bar{R}\bar{e}_k) \cong {}_R\bar{R}\bar{e}_t$. $\mathrm{Hom}({}_RM, {}_R\bar{R}\bar{e}_i) = 0$ if $i \neq k$, $\mathrm{Hom}(M_R, \bar{e}_t\bar{R}_R) \cong \bar{e}_k\bar{R}_R$. $\mathrm{Hom}(M_R, \bar{e}_i\bar{R}_R) = 0$ if $i \neq t$,

(iii) ${}_RM \cong {}_R(\bar{R}\bar{e}_k)^{n(t)}$, $M_R \cong (\bar{e}_t\bar{R})_R^{n(k)}$.

PROOF. (i) Since $\bar{e}_t$ is a primitive idempotent element in the endomorphism ring $\bar{R}_t$ of ${}_RM$, $Me_t = M\bar{e}_t$ is an indecomsable submodule of ${}_RM$. But ${}_RM$ and hence ${}_RMe_t$ is semisimple, so it follows that ${}_RMe_t$ is a simple submodule of ${}_RM$. Thus ${}_RMe_t \cong {}_R\bar{R}\bar{e}_k$. By left-right analogy, we can prove that $e_kM \cong \bar{e}_t\bar{R}_R$.

(ii) By Theorem 1.15, the functor $\mathrm{Hom}_{\bar{R}_k}(M, \ ) = \mathrm{Hom}({}_RM, \ )$ gives an equivalence $\bar{R}_k\text{-Mod} \longrightarrow \bar{R}_t\text{-Mod}$. In particular, since $\bar{R}\bar{e}_k$ is a simple left $\bar{R}_k$-module, the corresponding $\mathrm{Hom}({}_RM, {}_R\bar{R}\bar{e}_k)$ must be a simple left $\bar{R}_t$-module; but $\bar{R}\bar{e}_t$ is (up to isomorphism) the only simple left $\bar{R}\bar{e}_t$-module, so that we have ${}_R\mathrm{Hom}({}_RM, {}_R\bar{R}\bar{e}_k) \cong {}_R\bar{R}\bar{e}_t$. Let now $i \neq k$. Then that $\mathrm{Hom}({}_RM, {}_R\bar{R}\bar{e}_i) = 0$ follows from the fact that ${}_R\bar{R}\bar{e}_i \not\cong {}_R\bar{R}\bar{e}_k$ and ${}_RM$ is a direct sum of copies of ${}_R\bar{R}\bar{e}_k$. The other part can be proved in the similar way if we observe $\mathrm{Hom}_{\bar{R}_t}(M,\ )$ instead of $\mathrm{Hom}_{\bar{R}_k}(M,\ )$.

(iii) Consider again the equivalent functor $\mathrm{Hom}_{\bar{R}_k}(M,\ )$. Since ${}_R\mathrm{Hom}({}_RM, {}_R\bar{R}\bar{e}_k) \cong {}_R\bar{R}\bar{e}_t$ by (ii), we have ${}_R\mathrm{Hom}({}_RM, {}_R(\bar{R}\bar{e}_k)^{n(t)}) \cong {}_R\mathrm{Hom}({}_RM, {}_R\bar{R}\bar{e}_k)^{n(t)} \cong {}_R(\bar{R}\bar{e}_t)^{n(t)} \cong {}_R\bar{R}_t$. On the other hand, that $\bar{R}_t$ is the endomorphism ring of ${}_RM$ means that ${}_R\mathrm{Hom}({}_RM, {}_RM) \cong {}_R\bar{R}_t$. Thus it follows that ${}_R(\bar{R}\bar{e}_k)^{n(t)} \cong {}_RM$. Similarly, we have that $(\bar{e}_t\bar{R})_R^{n(k)} \cong M_R$ by considering $\mathrm{Hom}_{\bar{R}_t}(M,\ )$.

We see that the left artinian ring R with the two-sided composition series

$$R = I_0 > I_1 > \ldots > I_{s-1} > I_s = 0.$$

is exact if each simple two-sided R-module I_{p-1}/I_p is exact, i.e., a balanced bimodule. By the Jordan-Holder theorem (see Anderson-Fuller [74]), the composition factor module series $I_0/I_1, I_1/I_2, \ldots, I_{s-1}/I_s$ is, up to isomorphism and order, uniquely determined by R. Therefore, the notion of the exactness for R depends only on R and independent of the choice of the composition series. We now fix the above composition series once for all, and let $\bar{R}_{k(p)}$ and $\bar{R}_{t(p)}$ denote the left and the right simple components belonging to I_{p-1}/I_p, respectively. If R is exact, then the right R-module I_{p-1}/I_p is of finite length (indeed, its length is $n(k(p))$ by Lemma 13.1(iii)) for each p and consequently the right R-module R is of finite length, i.e., R is right artinian. Thus the concept of the exactness for R is left-right symmetric.

THEOREM 13.2. Let R be an exact ring. Then, for any indices p and i, ${}_R(I_{p-1}e_i/I_pe_i) \cong {}_R(\bar{R}\bar{e}_{k(p)})$ or $I_{p-1}e_i = I_pe_i$ according to $i = t(p)$ or $i \neq t(p)$. In particular, the series of left ideals

$$Re_i = I_0e_i > I_1e_i > \dots > I_{s-1}e_i > I_se_i = 0$$

gives a composition series of the left ideal Re_i if those terms I_pe_i for which $i \neq t(p)$ are deleted out of the series.

PROOF. If we observe that $I_{p-1}e_i \cap I_p = I_pe_i$, we have the isomorphism ${}_R(I_{p-1}e_i/I_pe_i) \cong {}_R((I_{p-1}e_i + I_p)/I_p) = {}_R(I_{p-1}/I_p)e_i$. Since I_{p-1}/I_p is an exact simple two-sided R-module, it follows from Lemma 13.1(i) that ${}_R(I_{p-1}/I_p)e_i \cong {}_R\bar{R}\bar{e}_{k(p)}$ is $i = t(p)$. On the other hand, since $\bar{e}_i \in \bar{R}_i$, we have $(I_{p-1}/I_p)e_i = (I_{p-1}/I_p)\bar{e}_i = 0$ if $i \neq t(p)$. This proves the theorem.

From Theorem 13.2 and its left-right analogy follows

COROLLARY 13.3. Let R be an exact ring. Then, for any indices i and j, the following are equal:

(a) The number of indices i such that $k(p) = i$ and $t(p) = j$;

(b) The multiplicity of the simple left R-module $\bar{R}\bar{e}_i$ in the composition factor module series of the left ideal Re_j;

(c) The multiplicity of the simple right R-module $\bar{e}_j\bar{R}$ in the composition factor module series of the right ideal e_iR.

Let $E_i = E({}_R\bar{R}\bar{e}_i)$, the injective envelope of the simple left R-module $\bar{R}\bar{e}_i$. The following results state that each E_i is of finite length.

THEOREM 13.4. Let R be an exact ring. For any indices p and i, ${}_R(\mathrm{Ran}_{E_i}(I_p)/\mathrm{Ran}_{E_i}(I_{p-1})) \cong {}_R\bar{R}\bar{e}_{t(p)}$ or $\mathrm{Ran}_{E_i}(I_p) = \mathrm{Ran}_{E_i}(I_{p-1})$ according to $i = k(p)$ or $i \neq k(p)$. In particular, the series

$$E_i = \mathrm{Ran}_{E_i}(I_s) > \mathrm{Ran}_{E_i}(I_{s-1}) > \dots > \mathrm{Ran}_{E_i}(I_1) > \mathrm{Ran}_{E_i}(I_0) = 0$$

gives a composition series of ${}_R E_i$ if those term $\mathrm{Ran}_{E_i}(I_p)$ for which $i \neq k(p)$ are deleted out of the series.

PROOF. Consider the exact sequence of two-sided R-modules

$$0 \longrightarrow {}_R(I_{p-1}/I_p)_R \longrightarrow {}_R(R/I_p)_R \longrightarrow {}_R(R/I_{p-1})_R \longrightarrow 0.$$

Since ${}_R E_i$ is injective, we have then an exact sequence of left R-modules

$$0 \longrightarrow {}_R\mathrm{Hom}_R(R/I_{p-1}, E_i) \longrightarrow {}_R\mathrm{Hom}_R(R/I_p, E_i)$$
$$\longrightarrow {}_R\mathrm{Hom}_R(I_{p-1}/I_p, E_i) \longrightarrow 0.$$

The second and the third terms of this sequence are naturally identified with $\mathrm{Ran}_{E_i}(I_{p-1})$ and $\mathrm{Ran}_{E_i}(I_p)$, respectively, while the fourth term coincides with $\mathrm{Hom}_R(I_{p-1}/I_p, \bar{R}\bar{e}_i)$ since ${}_R(I_{p-1}/I_p)$ is semisimple and $\bar{R}\bar{e}_i$ is the only simple submodule of ${}_R E_i$. Thus we have

$${}_R(\mathrm{Ran}_{E_i}(I_p)/\mathrm{Ran}_{E_i}(I_{p-1})) \cong {}_R\mathrm{Hom}_R(I_{p-1}/I_p, \bar{R}\bar{e}_i).$$

Since, however, I_{p-1}/I_p is an exact simple two-sided R-module, the right side of this isomorphism is isomorphic to ${}_R\bar{R}\bar{e}_{t(p)}$ or $= 0$ according to $i = k(p)$ or $i \neq k(p)$ by Lemma 13.1(ii). This completes the proof of the theorem.

The following is an immediate consequence of Theorem 13.4 and its left-right analogy, where $F_j = E(\bar{e}_j\bar{R}_R)$ is the injective envelope of the simple right R-module $\bar{e}_j\bar{R}$.

COROLLARY 13.5. Let R be an exact ring. Let i and j be any indices. Then the following are equal:

(a) The number of indices i such that $k(p) = i$ and $t(p) = j$;

(d) The multiplicity of the simple left R-module $\bar{R}\bar{e}_j$ in the composition factor module series of E_i;

(e) The multiplicity of the simple right R-module $\bar{e}_i\bar{R}$ in the composition factor module series of F_j.

REMARK 13.6. According to Corollaries 13.3 and 13.5, (a)-(e) given there are all equal if R is an exact ring, and R has a duality induced by ${}_R(\oplus_i E_i)$.

EXAMPLE 13.7. Every commutative artinian ring is exact. This is because if R is a commutative ring, then every simple R-module is isomorphic to the factor module R/I modulo a maximal ideal I of R and its endomorphism ring is the factor field R/I. It follows from Muller's Theorem (Theorem 4.8) that every commutative artinian ring has self-duality (see Corollary 6.9).

EXAMPLE 13.8. Every artinian semisimple ring R is exact. For, in this case, the simple componenets $R_1, R_2, \ldots, R_\ell$ of R form the two-sided composition factor module series of R and the endomorphism ring of each left R-module R_i is the simple ring R_i itself. A semisimple ring R has self-duality induced by ${}_RR_R$.

Azumaya's Conjecture: Every exact ring has self-duality.

SECTION 14. EXACT BIMODULES AND RINGS

If R is an exact ring and ${}_RE$ is a finitely cogenerated injective cogenerator, then ${}_RE$ defines a duality from the result in the above section. In this section, we shall show that $S = \mathrm{End}({}_RE)$ is still an exact ring, moreover R and S have the same upper and lower Loewy series.

Adapting Azumaya's exactness [83] to bimodules, Camillo, Fuller and Haack [86] said that a bimodule that has a composition series whose composition factors are balanced is an <u>exact bimodule</u> and that a ring R is an <u>exact ring</u> in case the regular bimodule ${}_RR_R$ is an exact bimodule. So Azumaya's exact rings are exact artinian rings in this sense. Like exact rings, exactness for modules is independent of the choice of the composition series. Using Azumaya's observation that exactness for artinian rings is left-right symmetric we see that a bimodule ${}_RM_S$ over semilocal rings R and S is exact if it has a composition series such that the left R-homomorphisms of each compsition factor is given by the right-multiplication of an element of an element of S.

The following result was proved by Habeb [89] and Xue [90].

THEOREM 14.1. Let R be an exact artinian ring and ${}_RE$ a finitely cogenerated injective cogenerator. Let $S = \mathrm{End}({}_RE)$. Then ${}_RE_S$ is an exact bimodule and S is an exact artinian ring.

PROOF. Let

$$R = I_0 > I_1 > \dots > I_{s-1} > I_s = 0.$$

be a two-sided composition series of R. So each I_{p-1}/I_p is an exact simple two-sided R-module. Let $\bar{R}_{k(p)}$ and $\bar{R}_{t(p)}$ denote the left and the right simple components belonging to I_{p-1}/I_p, respectively, as defined in Section 13. Since ${}_RE_S$ defines a duality, $(I_{p-1}/I_p)^* = \mathrm{Hom}_R(I_{p-1}/I_p, E)$ is a simple two-sided R-S-bimodule by Theorem 2.6. Using the method in the proof of Theorem 13.4, we have $\mathrm{Hom}_R(I_{p-1}/I_p, E) \cong \mathrm{Ran}_E(I^p)/\mathrm{Ran}_E(I^{p-1})$, so we have a composition series

$${}_RE_S = \mathrm{Ran}_E(I_s) > \mathrm{Ran}_E(I_{s-1}) > \dots > \mathrm{Ran}_E(I_1) > \mathrm{Ran}_E(I_0) = 0$$

of the two-sided R-S-bimodule E. Since I_{p-1}/I_p is a right $\bar{R}_{t(p)}$-module, $\mathrm{Ran}_E(I^p)/\mathrm{Ran}_E(I^{p-1})$ is a left $\bar{R}_{t(p)}$-module. Now S is right artinian and $\mathrm{Ran}_E(I^p)/\mathrm{Ran}_E(I^{p-1})$ is a simple R-S-bimodule, so there is a right simple component $\bar{S}_{u(p)}$ belonging to $\mathrm{Ran}_E(I^p)/\mathrm{Ran}_E(I^{p-1})$, i.e., $\mathrm{Ran}_E(I^p)/\mathrm{Ran}_E(I^{p-1})$ is a right $\bar{S}_{u(p)}$-module. Let f be an S-endomorphism of $\mathrm{Ran}_E(I^p)/\mathrm{Ran}_E(I^{p-1})$. Then $f^* = \mathrm{Hom}_S(f, E)$ is an R-endomorphism of $(\mathrm{Ran}_E(I^p)/\mathrm{Ran}_E(I^{p-1}))^* \cong I_{p-1}/I_p$. By the exactness of

I_{p-1}/I_p there exists an element $\bar{r} \in \bar{R}_{t(p)}$ such that $f^*(x) = x\bar{r}$ for all $x \in I_{p-1}/I_p$. Taking the duality again of f^*, we see that $f(y) = \bar{r}y$ for all $y \in \mathrm{Ran}_E(I^p)/\mathrm{Ran}_E(I^{p-1})$. So $\mathrm{Ran}_E(I^p)/\mathrm{Ran}_E(I^{p-1})$ is a simple balanced R-S-bimodule, and hence ${}_RE_S$ is exact. By the same argument we can show that S is an exact (artinian) ring by observing that ${}_RE$ and $({}_RE)^*_S = \mathrm{Hom}_R(E,E) = S_S$ form a duality pair with respect to E.

The following four lemmas will be used very often in order to prove that a basic exact artinian ring and the endomorphism ring of its minimal cogenerator have the same upper and lower Loewy series.

LEMMA 14.2. Let ${}_RM_S$ and ${}_RE$ be modules with S a semilocal ring. If M_S is semisimple, so is ${}_S\mathrm{Hom}_R(M, E)$.

PROOF. $MJ(S) = 0$ implies that $J(S)\mathrm{Hom}_R(M, E) = 0$.

From the proof of Theorem 13.4, the following result is immediate.

LEMMA 14.3. If ${}_RE_S$ defines a duality and $E_1 \leq E_2 \leq {}_RE_S$, then ${}_S\left[\mathrm{Ran}_S(E_1)/\mathrm{Ran}_S(E_2)\right]_S \cong \mathrm{Hom}_R(E_2/E_1, E)$ as bimodules.

LEMMA 14.4. Suppose that ${}_RE_S$ defines a duality. Let $J = J(R)$, and $N = J(S)$. Then

(1) $\mathrm{Ran}_E(J^i) = \mathrm{Lan}_E(N^i)$ for all $i \geq 0$;

(2) $\mathrm{Ran}_E\mathrm{Ran}_R(J^i) = \mathrm{Lan}_E\mathrm{Ran}_S(N^i)$ for all $i \geq 0$;

(3) $\mathrm{Ran}_E\mathrm{Lan}_R(J^i) = \mathrm{Lan}_E\mathrm{Lan}_S(N^i)$ for all $i \geq 0$.

PROOF. We use induction on i.

(1) It is obvious for $i = 0$. For $i = 1$, by Lemma 11.8 we have $\mathrm{Ran}_E(J) = \mathrm{Soc}({}_RE) = \mathrm{Soc}(E_S) = \mathrm{Lan}_E(N)$. Now let

$$\mathrm{Ran}_E(J^i) = \mathrm{Lan}_E(N^i) \quad \text{for some} \quad i \geq 1.$$

We consider $\mathrm{Ran}_E(J^{i+1})$ and $\mathrm{Lan}_E(N^{i+1})$. If $J^{i+1}x = 0$ for some $x \in E$, then $Jx \subseteq \mathrm{Ran}_E(J^i) = \mathrm{Lan}_E(N^i)$, and so $JxN^i = 0$, which implies that $xN^i \subseteq \mathrm{Ran}_E(J) = \mathrm{Lan}_E(N)$, and then $xN^{i+1} = (xN^i)N = 0$. Hence $\mathrm{Ran}_E(J^{i+1}) \subseteq \mathrm{Lan}_E(N^{i+1})$, and the other containment can be proved similarly. Therefore $\mathrm{Ran}_E(J^{i+1}) = \mathrm{Lan}_E(N^{i+1})$, and the proof is complete.

(2) First at all it clearly has

$$\mathrm{Ran}_E\mathrm{Ran}_R(J^0) = E = \mathrm{Lan}_E\mathrm{Ran}_S(N^0).$$

We assume that

$$\mathrm{Ran}_E\mathrm{Ran}_R(J^i) = \mathrm{Lan}_E\mathrm{Ran}_S(N^i) \quad \text{for some} \quad i \geq 1,$$

and examine $\mathrm{Ran}_E\mathrm{Ran}_R(J^{i+1})$ and $\mathrm{Lan}_E\mathrm{Ran}_S(N^{i+1})$.

Since ${}_R(\mathrm{Ran}_R(J^{i+1})/\mathrm{Ran}_R(J^i))$ is semisimple,

$$\left[\mathrm{Ran}_E\mathrm{Ran}_R(J^i)/\mathrm{Ran}_E\mathrm{Ran}_R(J^{i+1})\right]_S \cong \mathrm{Hom}_R\left[\mathrm{Ran}_R(J^{i+1})/\mathrm{Ran}_R(J^i),\ E\right]_S$$

is semisimple, and then

$$\begin{aligned} &{}_S\left[\mathrm{Ran}_S\mathrm{Ran}_E\mathrm{Ran}_R(J^{i+1})/\mathrm{Ran}_S\mathrm{Ran}_E\mathrm{Ran}_R(J^i)\right] \\ &\cong {}_S\mathrm{Hom}_R\left[\mathrm{Ran}_E\mathrm{Ran}_R(J^i)/\mathrm{Ran}_E\mathrm{Ran}_R(J^{i+1}), E\right] \end{aligned}$$

is semisimple, i.e., ${}_S\left[\mathrm{Ran}_S\mathrm{Ran}_E\mathrm{Ran}_R(J^{i+1})/\mathrm{Ran}_S(N^i)\right]$ is semisimple, since $\mathrm{Ran}_S\mathrm{Ran}_E\mathrm{Ran}_R(J^i) = \mathrm{Ran}_S(N^i)$ from $\mathrm{Ran}_E\mathrm{Ran}_R(J^i) = \mathrm{Lan}_E\mathrm{Ran}_S(N^i)$. Now since ${}_S\left[\mathrm{Ran}_S(N^{i+1})/\mathrm{Ran}_S(N^i)\right] = \mathrm{Soc}\left[{}_S(S/\mathrm{Ran}_S(N^i)\right]$, $\mathrm{Ran}_S\mathrm{Ran}_E\mathrm{Ran}_R(J^{i+1}) \subseteq \mathrm{Ran}_S(N^{i+1})$, and then $\mathrm{Ran}_E\mathrm{Ran}_R(J^{i+1}) = \mathrm{Lan}_E\mathrm{Ran}_S\mathrm{Ran}_E\mathrm{Ran}_R(J^{i+1}) \supseteq \mathrm{Lan}_E\mathrm{Ran}_R(N^{i+1})$. The other containment can be proved similarly, and we have

$$\mathrm{Ran}_E\mathrm{Ran}_R(J^{i+1}) = \mathrm{Lan}_E\mathrm{Ran}_S(N^{i+1}).$$

(3) By a similar proof as that given above.

The following useful result was observed by Camillo, Fuller, and Haack [86].

LEMMA 14.5. Let R and S be basic semiperfect rings. Then a simple bimodule ${}_RM_S$ is balanced if and only if ${}_RM$ and M_S are simple modules.

PROOF. Let $\bar{R} = R/\mathrm{Lan}_R(M)$ and $\bar{S} = S/\mathrm{Ran}_S(M)$.

(<==). Suppose that ${}_{\bar{R}}M$ and $M_{\bar{S}}$ are one diemnsional vector spaces. Let $T = \mathrm{End}({}_{\bar{R}}M)$. Then T is a division ring and $\mathrm{End}(M_T) \cong \bar{R}$ is too, so $M_T \cong T_T$. Now the right multiplication map $f : \bar{S} \longrightarrow T$ is an injective ring homomorphism and yields $M_{\bar{S}} \cong T_{\bar{S}}$. But then $\dim(T_{\bar{S}}) = \dim(M_{\bar{S}}) = 1$, so f is surjective and ${}_RM_S$ is balanced.

(==>). If $\mathrm{End}({}_RM) \cong \bar{S}$ and $\mathrm{End}(M_S) \cong \bar{R}$ are division rings then ${}_{\bar{R}}M$ and $M_{\bar{S}}$ must be (indecomposable and hence) one dimensional.

From now on through the end of this section, we make the following assumption

(#) R is a basic exact artinian ring with a basic set of primitive idempotents $e_1,\ldots,e_n$, $J = J(R)$, ${}_RE = \oplus_{j=1}^{n} E(Re_j/Je_j)$ the minimal injective cogenerator, $S = \mathrm{End}({}_RE)$ with $N = J(S)$, and $f_j : {}_RE \longrightarrow E(Re_j/Je_j))$ the jth projections.

Then ${}_RE_S$ defines a duality (Section 13), and ${}_RE_S$ is an exact bimodule and S is a basic exact artinian ring by Theorem 14.1. According to Corollaries 13.3 and 13.5 and Lemma 14.5, we have

$$c({}_RR) = c(R_R) = c({}_RE) = c(E_S) = c({}_SS) = c(S_S).$$

PROPOSITION 14.6. If

$${}_RR_R = I_0 > I_1 > \ldots > I_{s-1} > I_s = 0$$

is a composition series with

$$ {}_R(I_{i-1}/I_i) \cong Re_{j_i}/Je_{j_i} \quad \text{and} \quad (I_{i-1}/I_i)_R \cong e_{k_i}R/e_{k_i}J $$

then

$$ {}_S S_S = \mathrm{Ran}_S\mathrm{Ran}_E(I_0) > \mathrm{Ran}_S\mathrm{Ran}_E(I_1) > \dots $$
$$ \dots > \mathrm{Ran}_S\mathrm{Ran}_E(I_{n-1}) > \mathrm{Ran}_S\mathrm{Ran}_E(I_n) = 0 $$

is a composition series with

$$ {}_S\left[\mathrm{Ran}_S\mathrm{Ran}_E(I_{i-1})/\mathrm{Ran}_S\mathrm{Ran}_E(I_i)\right] \cong Sf_{j_i}/Nf_{j_i} $$

and

$$ \left[\mathrm{Ran}_S\mathrm{Ran}_E(I_{i-1})/\mathrm{Ran}_S\mathrm{Ran}_E(I_i)\right]_S \cong f_{k_i}S/f_{k_i}N. $$

PROOF. The exact bimodule ${}_RE_S$ has a composition series

$$ {}_RE_S = \mathrm{Ran}_E(I_n) > \mathrm{Ran}_E(I_{n-1}) > \dots > \mathrm{Ran}_E(I_1) > \mathrm{Ran}_E(I_0) = 0, $$

and then each $\mathrm{Ran}_E(I_i)/\mathrm{Ran}_E(I_{i-1})$ is simple on each side by Lemma 14.5. Moreover since

$$ {}_R\left[\mathrm{Ran}_E(I_i)/\mathrm{Ran}_E(I_{i-1})\right]_S \cong \mathrm{Hom}_R(I_{i-1}/I_i, E) $$
$$ = \mathrm{Hom}_R(I_{i-1}/I_i, E(Re_{j_i}/Je_{j_i}), $$

we have $\left[\mathrm{Ran}_E(I_i)/\mathrm{Ran}_E(I_{i-1})\right]_S \cong f_{j_i}S/N_{j_i}$, and ${}_R\left[\mathrm{Ran}_E(I_i)/\mathrm{Ran}_E(I_{i-1})\right] \cong Re_{k_i}/Je_{k_i}$ since $e_{k_i}\left[\mathrm{Ran}_E(I_i)/\mathrm{Ran}_E(I_{i-1})\right] \neq 0$. Now we have a composition series

$$ {}_S S_S = \mathrm{Ran}_S\mathrm{Ran}_E(I_0) > \mathrm{Ran}_S\mathrm{Ran}_E(I_1) > \dots $$
$$ \dots > \mathrm{Ran}_S\mathrm{Ran}_E(I_{n-1}) > \mathrm{Ran}_S\mathrm{Ran}_E(I_n) = 0 $$

and

$$ {}_S\left[\mathrm{Ran}_S\mathrm{Ran}_E(I_{i-1})/\mathrm{Ran}_S\mathrm{Ran}_E(I_i)\right]_S \cong \mathrm{Hom}_R\left[\mathrm{Ran}_E(I_i)/\mathrm{Ran}_E(I_{i-1}), E\right] $$
$$ = \mathrm{Hom}_R\left[\mathrm{Ran}_E(I_i)/\mathrm{Ran}_E(I_{i-1}), E(Re_{k_i}/Je_{k_i})\right], $$

and the results follow by the same arguments as above.

Two modules ${}_RM$ and ${}_SU$ are said to have the <u>same upper (lower) Loewy series</u> in case for each $i \geq 0$,

$$J^{i-1}M/J^iM \cong \oplus_j (Re_j/Je_j)^{k_{ij}}$$

if and only if

$$N^{i-1}U/N^iU \cong \oplus_j (Sf_j/Nf_j)^{k_{ij}}$$

$$\Big[\text{ resp. } \quad Ran_M(J^i)/Ran_M(J^{i-1}) \cong \oplus_j (Re_j/Je_j)^{l_{ij}}$$

if and only if

$$r_U(N^i)/r_U(N^{i-1}) \cong \oplus_j (Sf_j/Nf_j)^{l_{ij}} \qquad \Big].$$

Similar definition stands for right modules.

Azumaya [83] conjectured that exact artinian rings have self-duality, i.e., $R \cong S$ in our notation (#). We are unable to settle Azumaya's conjecture here, but the next result gives further evidence.

THEOREM 14.7 (Xue [90]). For each j, Re_j (resp. e_jR) and Sf_j (resp. f_jS) have the same upper and lower Loewy series. Consequently ${}_RR$ (resp. R_R) and ${}_SS$ (resp. S_S) have the same upper and lower Loewy series.

PROOF. We only prove the results for left modules. Let

$$R > \ldots > J^{i-1} = I_1 > I_2 > \ldots > I_{t-1} > I_t = J^i > \ldots > 0$$

be a composition series for ${}_RR_R$. Since each I_k/I_{k+1} is simple on each side by Lemma 14.5,

$$J^{i-1}/J^i \cong \oplus_{k=1}^{t-1} I_k/I_{k+1}$$

as left and right modules. By Theorem 13.2,

$$J^{i-1}e_j/J^ie_j \cong \oplus_{k=1}^{t-1} I_ke_j/I_{k+1}e_j$$

as left modules. Now by Proposition 14.6, we have a composition series

$$\begin{aligned} {}_SS_S > \ldots > \mathrm{Ran}_S\mathrm{Ran}_E(J^{i-1}) &= \mathrm{Ran}_S\mathrm{Ran}_E(I_1) > \ldots \\ \ldots > \mathrm{Ran}_S\mathrm{Ran}_E(I_t) &= \mathrm{Ran}_S\mathrm{Ran}_E(J^i) > \ldots > 0, \end{aligned}$$

and using Lemma 14.4(1) we have

$$\begin{aligned} N^{i-1}/N^i &= \mathrm{Ran}_S\mathrm{Lan}_E(N^{i-1})/\mathrm{Ran}_S\mathrm{Lan}_E(N^i) \\ &= \mathrm{Ran}_S\mathrm{Ran}_E(J^{i-1})/\mathrm{Ran}_S\mathrm{Ran}_E(J^i) \\ &\cong \oplus_{k=1}^{t-1}\left[\mathrm{Ran}_S\mathrm{Ran}_E(I_k)/\mathrm{Ran}_S\mathrm{Ran}_E(I_{k+1})\right] \end{aligned}$$

as left and right modules. And then using Theorem 13.2 again we have

$$N^{i-1}f_j/N^if_j \cong \oplus_{k=1}^{t-1}\left[\mathrm{Ran}_S\mathrm{Ran}_E(I_k)f_j/\mathrm{Ran}_S\mathrm{Ran}_E(I_{k+1})f_j\right]$$

as left modules. By Proposition 14.6 again

$$I_ke_j/I_{k+1}e_j \begin{cases} = 0 \\ \cong Re_m/Je_m \end{cases}$$

if and only if

$$\mathrm{Ran}_S\mathrm{Ran}_E(I_k)f_j/\mathrm{Ran}_S\mathrm{Ran}_E(I_{k+1})f_j \begin{cases} = 0 \\ \cong Sf_m/Nf_m \end{cases}$$

which completes the proof for the upper Loewy series. For the lower Loewy series, let

$$R > \ldots > \mathrm{Ran}_R(J^i) = I_1 > I_2 > \ldots > I_{t-1} > I_t = \mathrm{Ran}_R(J^{i-1}) > \ldots > 0$$

be a composition series for ${}_RR_R$.

Since each I_k/I_{k+1} is simple on each side

$$\mathrm{Ran}_R(J^i)/\mathrm{Ran}_R(J^{i-1}) \cong \oplus_{k=1}^{t-1} I_k/I_{k+1}$$

as left modules. By Theorem 13.2,

$$\mathrm{Ran}_{Re_j}(J^i)/\mathrm{Ran}_{Re_j}(J^{i-1}) = \mathrm{Ran}_R(J^i)e_j/\mathrm{Ran}_R(J^{i-1})e_j$$
$$\cong \oplus_{k=1}^{t-1} I_k e_j/I_{k+1}e_j$$

as left modules. By Proposition 14.6, we have a composition series for ${}_S S_S$:

$$S > \ldots > \mathrm{Ran}_S\mathrm{Ran}_E\mathrm{Ran}_R(J^i) = \mathrm{Ran}_S\mathrm{Ran}_E(I_1) > \ldots$$
$$\ldots > \mathrm{Ran}_S\mathrm{Ran}_E(I_t) = \mathrm{Ran}_S\mathrm{Ran}_E\mathrm{Ran}_R(J^{i-1}) > \ldots > 0.$$

Using Lemma 14.4(2), we have

$$\mathrm{Ran}_S(N^i)/\mathrm{Ran}_S(N^{i-1}) = \mathrm{Ran}_S\mathrm{Ran}_E\mathrm{Ran}_R(J^i)/\mathrm{Ran}_S\mathrm{Ran}_E\mathrm{Ran}_R(J^{i-1})$$
$$\cong \oplus_{k=1}^{t-1}\left[\mathrm{Ran}_S\mathrm{Ran}_E(I_k)/\mathrm{Ran}_S\mathrm{Ran}_E(I_{k+1})\right]$$

as left modules. By Theorem 13.2 again,

$$\mathrm{Ran}_{Sf_j}(N^i)/\mathrm{Ran}_{Sf_j}(N^{i-1}) = \mathrm{Ran}_S(N^i)f_j/\mathrm{Ran}_S(N^{i-1})f_j$$
$$\cong \mathrm{Ran}_S\mathrm{Ran}_E(I_k)f_j/\mathrm{Ran}_S\mathrm{Ran}_E(I_{k+1})f_j$$

as left modules. By Proposition 14.6 again, we get

$$I_k e_j/I_{k+1}e_j \begin{cases} = 0 \\ \cong Re_m/Je_m \end{cases}$$

if and only if

$$\mathrm{Ran}_S\mathrm{Ran}_E(I_k)f_j/\mathrm{Ran}_S\mathrm{Ran}_E(I_{k+1})f_j \begin{cases} = 0 \\ \cong Sf_m/Nf_m \end{cases}$$

which completes the proof of the theorem.

SECTION 15. LOCALLY DISTRIBUTIVE RINGS

A module M is called <u>distributive</u> in case its lattice $L(M)$ of submodules is distributive. That is, for all $A,B,C \in L(M)$,

$$A \cap (B + C) = (A \cap B) + (A \cap C)$$

or equivalently for all $A,B,C \in L(M)$,

$$A + (B \cap C) = (A + B) \cap (A + C).$$

Clearly submodules and quotient modules of a distributive module are distributive. An artinian ring R is called <u>locally distributive</u> in case each of its left and right indecomposable projective R-modules is distributive. In this section we show that locally distributive rings are exact rings (Camillo, Fuller and Haack [86]) and the endomorphism ring of a finitely cogenerated injective cogenerator over a locally distributive ring is also locally distributive (Fuller and Xue [91]). This lends support to Azumaya's conjecture that exact rings have self-duality.

We begin with a lemma that is adapted from Tachikawa [73].

LEMMA 15.1. Let F: R-Mod $\longrightarrow$ S-Mod define an equivalence. If ${}_RM_T$ is a balanced bimodule, then ${}_SFM_T$ is also balanced.

PROOF. There is a bimodule ${}_SP_R$ such that ${}_SP$ and P_R are progenerators and $FX = P\otimes_R X$ for every $X \in$ R-Mod. Now $T = \mathrm{End}({}_RM) = \mathrm{End}({}_SFM)$. There is an epimorphism

$$P_R \longrightarrow \mathrm{Hom}_T({}_RM_T,\ P\otimes_R M_T)$$

given by $p \longmapsto (m \longmapsto p\otimes m)$ for $p \in P$ and $m \in M$. Namely, such an epimorphism exists for $P_R = R_R$, since ${}_RM_T$ is balanced, and therefore also for a direct summand of a finite direct sum of copies of R_R, hence for a progenerator P_R. Then we get an epimorphism

$$\mathrm{Hom}_R(P_R,\ P_R) \longrightarrow \mathrm{Hom}_R(P_R,\ \mathrm{Hom}_T(M_T,\ P\otimes_R M_T)),$$

but the first group is just S, whereas the second is canonically isomorphic to

$$\mathrm{Hom}_T(P\otimes_R M_T,\ P\otimes_R M_T) \cong \mathrm{Hom}_T(FM,\ FM).$$

It can be checked easily that this surjection of S onto $\mathrm{End}({}_TFM)$ is the canonical one.

Under the assumption of the above lemma, if ${}_RM_T$ is also simple then ${}_RFM_T$ is simple, too, since the equivalence F induces an isomorphism between the lattices of submodules of ${}_RM$ and ${}_SFM$ under which R-T-submodule of M correspond to S-T-submodules of FM. Hence we have the following

LEMMA 15.2. Let F: R-Mod $\longrightarrow$ S-Mod define an equivalence. If ${}_RM_T$ is an exact bimodule, then ${}_SFM_T$ is also exact.

Now it is a simple matter to show that exactness of rings is preserved under Morita equivalence. This is due to Camillo, Fuller and Haack [86].

THEOREM 15.3. If R is an exact ring then so is every ring that is Morita equivalent to R.

PROOF. Suppose that S is equivalent to R. Then there is a balanced bimodule ${}_SP_R$ (with ${}_SP$ and P_R progenerators) such that $({}_SP\otimes_R-)$: R-Mod $\longrightarrow$ S-Mod and $\mathrm{Hom}_R({}_SP_R, -)$: Mod-R $\longrightarrow$ Mod-S are category equivalences. If R is an exact ring then by Lemma 15.2, ${}_SP_R \cong {}_SP\otimes_RR$ is exact and so is ${}_SS_S \cong \mathrm{Hom}_R({}_SP_R, {}_SP_R)$.

We need the folowing lemma due to Fuller [70] to characterize exact rings in terms of their indecomposable projective modules (see Anderson-Fuller [74]).

LEMMA 15.4. Let the left R-module M be the direct sum $M = M_1\oplus M_2$ of submodules M_1 and M_2. Then the restriction map Res is a ring homomorphism making the diagram

$$\begin{array}{ccc} & R & \\ {}^{\lambda}\swarrow & & \searrow^{\lambda} \\ \mathrm{BiEnd}({}_RM) & \xrightarrow{\ \mathrm{Res}\ } & \mathrm{BiEnd}({}_RM_1) \end{array}$$

commute, where $\mathrm{BiEnd}({}_RM) = \mathrm{End}(M_{\mathrm{End}({}_RM)})$. Moreover, if M_1 generates and cogenerates M, then Res is an isomorphism.

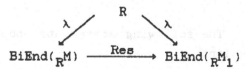

LEMMA 15.5. Let ${}_RM_S$ be an exact bimodule over semilocal rings R and S. If e is an idempotent in S such that $Me \neq 0$ then ${}_RMe_{eSe}$ is an exact bimodule.

PROOF. Let

$$ {}_RM_S = M_0 > M_1 > \ldots > M_n = 0 $$

be a composition series for ${}_RM_S$ and suppose that e is an idempotent in S with $Me \neq 0$. Then choosing $i_0 < \ldots < i_k$ so that $M_{i_j}e$ are the distinct members of $M_1e, \ldots, M_ne$ we obtain a composition series

$$ {}_RMe_{eSe} = M_{i_0}e > M_{i_1}e > \ldots > M_{i_k}e = 0 $$

for ${}_RMe_{eSe}$ with each

$$ M_{i_{j-1}}e/M_{i_j}e \cong (M_{i_{j-1}}/M_{i_j})e $$

as R-eSe bimodules. Thus we need only show that if ${}_RM_S$ is simple and balanced then the simple bimodule ${}_RMe_{eSe}$ is also balanced. If ${}_RM_S$ is balanced then multiplication by the elements of S comprises the endomorphism ring of ${}_RM$ so ${}_RM$ is a balanced module (i.e., the bimodule ${}_RM_{\mathrm{End}({}_RM)}$ is balanced); and if ${}_RM_S$ is simple then $MeS = M$ and $\mathrm{Lan}_M(Se) = 0$, i.e., the direct summand ${}_RMe$ of ${}_RM$ generates and cogenerates ${}_RM$, so ${}_RMe$ is balanced by lemma 15.4. Therefore, if ${}_RM_S$ is simple and balanced then so is ${}_RMe_{eSe}$.

In the following theorem, (a) ==> (b) is due to Azumaya [83] and the rest is due to Camillo, Fuller and Haack [86].

THEOREM 15.6. The following statements about an artinian ring R are equivalent:

(a) R is exact;

(b) Every left and every right indecomposable projective

R-module has a composition series whose terms are stable under endomorphisms;

(c) ${}_R Re_{eRe}$ and ${}_{eRe} eR_R$ are exact bimodules for every primitive idempotent e in R.

PROOF. Assume that R is basic.

(a) ==> (c). By lemma 15.5.

(c) ==> (b). If ${}_R Re_{eRe}$ is exact, then by Lemma 14.5 a composition series for this bimodule must have factors that are left simple, and so must be a composition series for ${}_R Re$ with terms stable under endomorphisms.

(b) ==> (a). If e is a primitive idempotent in R then every composition factor of the bimodule ${}_R Re_{eRe}$ is simple on the left. But since R is basic, if $I < K$ are ideals of R such that K/I is a composition factor of ${}_R R_R$, there is a primitive idempotent e in R with

$$K/I = (K/I)e \cong Ke/Ie$$

and the later is a composition factor of ${}_R Re_{eRe}$. Thus K/I is left, and similarly right, simple over R, so R is exact by Lemma 14.5.

The following three results are due to Stephenson [74]. The first lemma characterizes distributive modules by the homomorphisms into and out of them.

LEMMA 15.7. For a module M the following assertions are equivalent:

(1) M is distributive;

(2) For every module P and $f \in \mathrm{Hom}(P,M)$, $f^{-1}(A + B) = f^{-1}(A) + f^{-1}(B)$ for all $A, B \in L(M)$;

(3) For every module Q and $g \in \mathrm{Hom}(M,Q)$, $g(A \cap B) = g(a) \cap g(B)$ for all $A, B \in L(M)$.

PROOF. (2) ==> (1). Let C be any submodule of M and let

$f : C \longrightarrow M$ be the inclusion map. Then
$(A + B) \cap C = f^{-1}(A + B) = f^{-1}(A) + f^{-1}(B) = (A\cap B) + (B\cap C)$.

(1) ==> (2).

$$\begin{aligned} f^{-1}(A) + f^{-1}(B) &= f^{-1}[f(f^{-1}(A))] + f^{-1}[f(f^{-1}(B))] \\ &= f^{-1}[f(f^{-1}(A)) + f(f^{-1}(B))] \\ &= f^{-1}[(A \cap f(P)) + (B \cap f(P))] \\ &= f^{-1}[(A + B) \cap f(P)] \\ &= f^{-1}(A + B). \end{aligned}$$

(1) <==> (3). This is proved dually.

LEMMA 15.8. Let M be a distributive module with a submodule X. If $f,g \in \mathrm{Hom}(X,M)$, then
$X = g^{-1}(\mathrm{Im}(f)) + f^{-1}(\mathrm{Im}(g))$.

PROOF. Since $\mathrm{Im}(f + g) \subseteq \mathrm{Im}(f) + \mathrm{Im}(g)$, we have

$$\begin{aligned} X &= (f + g)^{-1}(\mathrm{Im}(f) + \mathrm{Im}(g)) \\ &= (f + g)^{-1}(\mathrm{Im}(f)) + (f + g)^{-1}(\mathrm{Im}(g)) \qquad \text{By Lemma 15.7} \\ &= g^{-1}(\mathrm{Im}(f)) + f^{-1}(\mathrm{Im}(g)). \end{aligned}$$

The next result states that submodules of a distributive module of finite length are stable under its endomorphisms. Stephenson [74] has a more general result than this.

PROPOSITION 15.9. Let M be a distributive module of finite length. If N is a submodule of M and $f \in \mathrm{End}(M)$, then $f(N) \subseteq N$.

PROOF. Let $A = \Sigma \{ B \leqq N \mid f(B) \subseteq B \}$. Clearly A is the largest submodule of N with $f(A) \subseteq A$.

Suppose that $f(N)$ is not contained in N, then $A \neq N$. Since M has finite length, we can find a submodule X such that $A \subseteq X \subseteq N$ with $A \neq X$ and that X/A is simple. But $A \subseteq f^{-1}(A) \subseteq f^{-1}(X)$, and so $A \subseteq X \cap f^{-1}(X) \subseteq X$. As $f(X)$ is not contained in X and X/A is simple, we get $A = X \cap f^{-1}(X)$.

Let $i : X \longrightarrow M$ be the inclusion map and let $g = f|_X$. Then by Lemma 15.8,

$$\begin{aligned} X &= g^{-1}(\mathrm{Im}(i)) + i^{-1}(\mathrm{Im}(g)) \\ &= (X \cap f^{-1}(X)) + i^{-1}(f(X)) = A + i^{-1}(f(X)). \end{aligned}$$

Hence $X = i(X) = i(A) + f(X) = A + f(X)$, and it follows that $f(X) \subseteq X$. This is a contradiction.

The following theorem follows now from Proposition 15.9 and Theorem 15.6.

THEOREM 15.10. Every locally distributive ring is an exact ring.

To prove that the endomorphism ring of the minimal left cogenerator over a locally distributive ring is still a locally distributive ring, we shall need a few results concerning distributive modules.

For elements x and y in a left R-module ${}_RM$ we define

$$(Rx : y) = \{\, r \in R \mid ry \in Rx \,\}.$$

The following is due to Stephenson [74].

THEOREM 15.11. The following are equivalent for a left R-module ${}_RM$:

(1) M is distributive;

(2) $(Ra : b) + (Rb : a) = R$ for all $a,b \in M$;

(3) $R(a+b) = (Ra \cap R(a+b)) + (Rb \cap R(a+b))$ for all $a,b \in M$;

(4) $Ra + Rb = R(a+b) + (Ra \cap Rb)$ for all $a,b \in M$.

PROOF. (1) ==> (4). $Ra \subseteq Ra + Rb = Rb + R(a+b)$. Hence, as M is distributive, we have $Ra = (Ra \cap Rb) + (Ra \cap R(a+b))$ and, similarly, $Rb = (Ra \cap Rb) + (Rb \cap R(a+b))$. Therefore

$$Ra + Rb = (Ra \cap Rb) + (Ra \cap R(a+b)) + (Rb \cap R(a+b))$$

$$= (Ra \cap Rb) + ((Ra + Rb) \cap R(a+b))$$
$$= (Ra \cap Rb) + R(a+b).$$

(4) ==> (2).

$$Ra = Ra \cap (Ra + Rb) = Ra \cap (R(a+b) + (Ra \cap Rb))$$
$$= (Ra \cap R(a+b)) + (Ra \cap Rb)$$
$$= (Ra : (a+b))(a+b) + (Ra \cap Rb)$$
$$= (Ra : b)(a+b) + (Ra \cap Rb)$$
$$= (Ra : b)a + (Ra \cap Rb) \quad \text{since} \quad (Ra : b)b = Ra \cap Rb,$$
$$= ((Ra : b) + (Rb : a))a.$$

Hence $R = (Ra : b) + (Rb : a)$ since $\mathrm{Lan}_R(a) \leq (Rb : a)$.

(2) ==> (3).

$$R(a+b) = ((Ra : b) + (Rb : a))(a+b)$$
$$= (Ra : b)(a+b) + (Rb : a)(a+b)$$
$$= (Ra : (a+b))(a+b) + (Rb : (a+b))(a+b)$$
$$= (Ra \cap R(a+b)) + (Rb \cap R(a+b))$$

(3) ==> (1). Let $A,B,C \in L(M)$ and suppose that $c = a+b \in C \cap (A+B)$, where $a \in A$, $b \in B$, and $c \in C$. By hypothesis, $Rc = R(a+b) = (Ra \cap Rc)+ (Rb \cap Rc)$, so $c \in (C \cap A) + (C \cap B)$. Therefore $C \cap (A + B) = (C \cap A) + (C \cap B)$ and M is distributive.

It is a simple matter to note that $M \oplus M$ never be distributive for any non-zero module M: take $A = M \oplus 0$, $B = 0 \oplus M$, and $C = \{(m,m) \mid m \in M\}$. Then $A \cap (B + C) = A$ but $(A \cap B) + (A \cap C) = 0$. A module has square-free socle if its socle has at most one copy of each simple module. Camillo [75] characterized distributive modules in terms of the socles of their quotients.

THEOREM 15.12. A left R-module ${}_RM$ is distributive if and only if for every submodule N, M/N has square-free socle.

PROOF. (==>). Every quotient and submodule of a distributive module are distributive. If M/N contains a submodule of the form $S \oplus S$, where S is simple, it will be distributive. This contradicts the fact that we established preceeding the theorem.

(<==). Using Theorem 15.11, we show that $R = (Ra : b) + (Rb : a)$ for all $a, b \in M$. We prove that this sum lies outside of every maximal left ideal of R. Let K be any such. Consider the module $(Ra + Rb)/(Ka + Kb)$. The images of a and b in this module generate simple or zero submodules, but if they both generated distinct simple submodules we would contradict the hypothesis. The above may be summarized by the statement that one of the following conditions must hold:

$$(1)\ a \in Rb + Ka + Kb = Ka + Rb;$$

$$(2)\ b \in Ra + Ka + Kb = Ra + Kb.$$

If (1) holds, write $a = rb + ka$ or $(1-k)a = rb$ so $1-k \in (Rb : a)$. Since $1-k \notin K$, $(Rb : a)$ is not contained in K. If (2) holds, then $(Ra : b)$ is not contained in K.

A module is called <u>uniserial</u> in case its lattice of submodules is a finite chain, i.e., any two submodules are comparable. It is not difficult to prove that the following statements about a module $M \neq 0$ over a semiprimary are equivalent: (1) M is uniserial; (2) M has a unique composition series; (3) The upper Loewy series of M is a composition series for M; (4) The lower Loewy series is a composition series for M.

The following characterization of distributive modules over semiperfect rings is given in Fuller [78].

LEMMA 15.13. The following statements about a left module M over a semiperfect ring R are equivalent:

(1) M is distributive;

(2) For each primitive idempotent e in R, the set of submodules $\{Rex \mid x \in M\}$ is linearly ordered;

(3) For each primitive idempotent e in R, the left eRe-module eM is uniserial.

PROOF. Let $J = J(R)$.

(1) ==> (2). Suppose $x, y \in M$ such that Rex and Rey are not comporable. Then, since Jex and Jey are their unique maximal submodules, $Rex \cap Rey = Jex \cap Jey$. It follows that

$$Re/Je \oplus Re/Je \cong (Rex + Rey)/(Jex + Jey)$$

which precludes the distributivity of M.

(2) ==> (3). If (2) holds then for all $ex, ey \in eM$ we must have $ex \in eRey$ or $ey \in eRex$, and so eM is uniserial.

(3) ==> (1). If M is not distributive, then by Theorem 15.12 there is a simple module Re/Je such that $(Re/Je \oplus Re/Je)$ embeds in M/N for some $N \leq M$. It follows that $e(M/N) \cong eM/eN$ is not uniserial.

The following three results are due to Fuller [69].

LEMMA 15.14. Let e be a primitive idempotent in a semiprimary ring R. Let $E = E(Re/Je)$ and suppose I is an ideal of R with $e \not\in I$. Then Re/Je is a simple R/I-module and $E({}_{R/I}Re/Je) \cong \mathrm{Ran}_E(I)$.

PROOF. If $I(Re/Je) \neq 0$ then Ie is not contained in Je, so $Ie = Re$ and $e \in I$. Thus by assumption $I(Re/Je) = 0$ and Re/Je is an R/I-module. Since any of its R/I-submodules must be R-submodules, Re/Je is simple over R/I. Also $Re/Je \subseteq \mathrm{Ran}_E(I)$ so $\mathrm{Ran}_E(I)$ is an essential extension of Re/Je. Suppose $N \leq {}_{R/I}M$ and $g : N \longrightarrow \mathrm{Ran}_E(I)$ is a map over R/I. Then we can regard g as an R-map $g : N \longrightarrow E$, so there is an R-map $\bar{g} : M \longrightarrow E$ that extends g. But $I\bar{g}(M) = \bar{g}(IM) = 0$ and we really have $\bar{g} : M \longrightarrow \mathrm{Ran}_E(I)$. This proves the lemma.

THEOREM 15.15. Let R be a left artinian ring with primitive idempotents e and f. Suppose there exist ideals I and K of R such that $\mathrm{Ran}_{E(Rf/Jf)}(I) \cong Re/Ke$. Then $fR/fI \cong \mathrm{Lan}_{E(eR/eJ)}(K)$.

PROOF. In the proof we shall use Lemma 11.4, its right-left

dual version and Lemma 11.10 without reference. Let e and f be primitive idempotents of R, let $E = E(Rf/Jf)$ and $E' = E(eR/eJ)$, and suppose $\mathrm{Ran}_E(I) \cong Re/Ke$ for some ideals I and K of R. Then we must have $I(Re/Ke) = 0$ and $K(\mathrm{Ran}_E(I)) = 0$ so that $Ie = IRe \leq Ke$ and $fK \leq \mathrm{Lan}_{fR}(\mathrm{Ran}_E(I)) = \mathrm{Lan}_{fR}(\mathrm{Ran}_E(fI)) = fI$. Thus

$$(I + K)e = Ke \text{ and } f(I + K) = fI.$$

Moreover we have

$$\mathrm{Ran}_E(I + K) = \mathrm{Ran}_E(f(I + K)) = \mathrm{Ran}_E(fI) = \mathrm{Ran}_E(I)$$

and

$$\mathrm{Lan}_{E'}(I + K) = \mathrm{Lan}_{E'}((I + K)e) = \mathrm{Lan}_{E'}(Ke) = \mathrm{Lan}_{E'}(K).$$

If $e \in I + K$ then $Re \leq (I + K)e = Ke$, $\mathrm{Ran}_E(fI) = \mathrm{Ran}_E(I) = 0$ and $fI = \mathrm{Lan}_{fR}(\mathrm{Ran}_E(fI)) = \mathrm{Lan}_{fR}(0) = fR$. So that

$$fR/fI \cong 0 = \mathrm{Lan}_{E'}(Re) = \mathrm{Lan}_{E'}(Ke) = \mathrm{Lan}_{E'}(K).$$

If $f \in I + K$ then $fR \leq f(I + K) = fI$, $\mathrm{Ran}_E(I) = \mathrm{Ran}_E(fI) = \mathrm{Ran}_E(fR) = 0$ and $Re = Ke$. So again

$$fR/fI \cong 0 = \mathrm{Lan}_{E'}(K).$$

If neither e nor f is contained in $I + K$ let

$$\bar{R} = R/(I+K),\ \bar{J} = J(\bar{R}),\ \bar{e} = e + (I+K), \text{ and } \bar{f} = f + (I+K).$$

Then as $\bar{R}$-modules

$$\bar{R}\bar{e} \cong Re/Ke,\ \bar{f}\bar{R} \cong fR/fI$$

and by Lemma 15.14 and its right-left symmetric version,

$$E(\bar{R}\bar{f}/\bar{J}\bar{f}) \cong \mathrm{Ran}_E(I),\ E(\bar{e}\bar{R}/\bar{e}\bar{J}) \cong \mathrm{Lan}_{E'}(K).$$

But now we have, over $\bar{R}$,

$$\bar{R}\bar{e} \cong E(\bar{R}\bar{f}/\bar{J}\bar{f}).$$

So, according to Theorem 11.13, $\mathrm{Soc}(\bar{f}\bar{R}) \cong \bar{e}\bar{R}/\bar{e}\bar{J}$ and $\bar{f}\bar{R}$ is injective over $\bar{R}$. That is, as $\bar{R}$-modules

$$\bar{f}\bar{R} \cong E(\bar{e}\bar{R}/\bar{e}\bar{J}).$$

Therefore

$$fR/fI \cong \mathrm{Lan}_{E^*}(K)$$

over R.

An artinian ring R is called <u>serial</u> (generalized uniserial in the sense of Nakayama [41]) in case each of its left and right indecomposable projective modules is uniserial. Clearly, serial rings are locally distributive. Amdal and Ringdal [68] claimed that serial rings have self-duality. This has not been proved until recently (see Dischinger and Müller [84] and Waschbüsch [86]). The proofs are so technical that we choose not to include here.

Theorem 15.16 is an interesting characterization of serial rings, which is due to Fuller [69]. We need a sublemma (Nakayama [40]) to prove this theorem.

SUBLEMMA. If R is a left artinian ring such that $R/J(R)^2$ is serial, then R is serial.

PROOF. Let $J = J(R)$ and e an primitive idempotem of R. Assume inductively, for $k \geq 1$, that $J^k e/J^{k+1}e$ is simple, so there is a primitive idempoten f in R such that we have a projective cover

$$Rf \longrightarrow J^k e \longrightarrow 0.$$

Then $Jf/J^2 f \cong J^{k+1}e/^{k+2}e$ unless the latter is 0.

THEOREM 15.16 (Fuller). A left artinian ring R is a serial ring if and only if the indecomposable projective left R-modules and injective left R-modules are uniserial.

PROOF. Let $J = J(R)$.

(<==). If ${}_R T$ is a simple R-module with $E = E(T)$, then since $JT = 0$, the injective envelope of T over R/J^2 is $\mathrm{Ran}_E(J^2)$ by Proposition 1.10. Of course, the projectives over

R/J^2 are factors of those over R. Thus by the Sublemma we may assume that $J^2 = 0$. Let f be a primitive idempotent in R and $fJ \neq 0$. Then the semisimple left R-module J contains a copy of Rf/Jf, so there is a non-zero homomorphism from J to $E = E(Rf/Jf)$ which must be multiplication by an element of E. Hence $JE \neq 0$, and then $c(E) = 2$ since ${}_RE$ is uniserial. If $E/JE \cong Re/Je$ for some primitive idempotent e, then $E \cong Re$. By Theorem 11.13 fR is injective with $Soc(fR) = fJ \cong eR/eJ$ and fR is uniserial.

(==>). Let e be a primitive idempotent of R and $E = E(Re/Je)$. We want to show that E is uniserial. Let k be the positive interger such that $J^kE = 0$ but $J^{k-1}E \neq 0$. Since E is the injective envelope of Re/Je over R/J^k which is still a serial ring, we may assume that $J^k = 0$.

Let $J^{k-1}fE \neq 0$ for some primitive idempotent $f \in R$, and let $J^{k-1}fx \neq 0$ for some $x \in E$. Since $J^k = 0$, we have

$$Soc(Rf) = J^{k-1}f \cong J^{k-1}fx = Soc(E) \cong Re/Je.$$

Hence Rf can be embedded into E. The proof will be completed if we prove that Rf is injective, since E is indecomposable. Since $Soc(Rf)J = J^{k-1}JJ = 0$, hence $Soc(Rf) \leq Lan_R(J) = Soc(R_R)$. Now

$$0 \neq e(Soc\ Rf) \leq e(Soc\ R_R)f = Soc(eR_R)f.$$

We know $Soc(eR)$ is simple, then we must have $Soc(eR) \cong fR/fJ$. By Theorem 11.13, both ${}_RRf$ and eR_R are injective.

An anlogous characterization of locally distributive rings is given as follows, where the necessary part is due to Fuller [78] and the sufficient one is due to Fuller and Xue [91]. We shall subsequently apply this result to examine the endomorphism rings of their minimal cogenerators.

THEOREM 15.17. A left artinian ring R is locally distributive if and only if its indecomposable left projective modules and injective modules are distributive.

PROOF. Let $J = J(R)$.

($\Longrightarrow$). Let e and f be primitive idempotents in R and let $E = E(Rf/Jf)$. Assuming the hypothesis, we shall first show that if $eE \neq 0$ then ReE/JeE is simple. Since $\mathrm{Ran}_E(fR) = 0$ by Lemma 11.4, we get $f \notin \mathrm{Lan}_R(ReE)$. Now by Lemma 11.10,

$$0 \neq ReE = \mathrm{Ran}_E\mathrm{Lan}_{fR}(ReE) = \mathrm{Ran}_E\mathrm{Lan}_R(ReE).$$

So by Lemma 15.14, ReE is the injective envelope of Rf/Jf over a factor ring of R, namely $R/\mathrm{Lan}_R(ReE)$. Thus we may assume that $ReE = E$, and hence that E/JE is e-homogeneous (i.e., a direct sum of copies of Re/Je). Further, since all the conditions involved are Morita invariants we may assume that R is a basic ring, so that if a semisimple module ${}_RM$ is e-homogeneous then $M = eM$. Having made these assumptions, let $S = \mathrm{Soc}(R_R) = \mathrm{Lan}_R(J)$. Then as in the proof of Theorem 11.12, $JE = \mathrm{Ran}_E(\mathrm{Lan}_R(J)) = \mathrm{Ran}_E(S)$, and by Lemma 11.5(2), $E/JE \cong \mathrm{Hom}_{fRf}(fS, fE)$ as left R-modules. Since E/JE is left e-homogeneous, $fS = \mathrm{Soc}(fR_R)$ is right e-homogeneous by Theorem 11.12. Thus, since fR is distributive, $fS \cong eR/eJ$. Since R is basic, $fS = fSe$; and by Lemma 15.13, fS is uniserial over fRf. But $fJfSe$ is a proper right submodule of fSe, so $fJfSe = 0$. Hence the bimodule ${}_{fRf}fSe_{eRe}$ is simple on each side. Since it is simple on the left, the fRf-maps $fS \longrightarrow fE$ have their images in $\mathrm{Soc}({}_{fRf}fE) \cong fRf/fJf \cong fSe$. (See Lemma 11.6.) Thus we have

$$\begin{aligned} e(E/JE) = E/JE &\cong \mathrm{Hom}_{fRf}(fS, fE) \\ &\cong \mathrm{Hom}_{fRf}(fSe, fSe) \\ &= \mathrm{Hom}_{fRf/fJf}(fSe_{eRe/eJe}, fSe) \end{aligned}$$

which, since fSe is an fRf/fJf-eRe/eJe bi-vector space that is 1-dimensional on each side, is 1-dimensional over eRe/eJe. This proves that, in the assumed setting, E/JE is simple. Therefore the hypothesis of the theorem implies that ReE/JeE is simple or

0 for each primitive idempotents e in R, and hence that ReE is an epimorph of Re and so is distributive. But then by Lemma 15.13 each ${}_{eRe}eE$ is uniserial amd E is distributive.

For sufficiency, let e and f be primitive idempotents in a ring R that satisfies the condition, and let $E = E(Rf/Jf)$. Then fR and E form a pair. We shall prove this implication by using Lemma 15.13 to show that fRe_{eRe} is uniserial. According to Lemma 11.6, $eE(Re/Je) = E(eRe/eJe)$, the left injective envelope of eRe/eJe. Since Re and E are distributive R-modules, eRe and eE are uniserial left eRe-modules by Lemma 15.13. Hence eRe is a local serial ring by Theorem 15.16. Thus it only remains to show that $fRe/fReJe$ is simple (or zero) over eRe. According to Lemma 11.4(1) and Lemma 11.10(2), for any ideal $I \leqq {}_RR_R$ and bisubmodule $Q \leqq {}_RE_{End({}_RE)}$ the annihilator conditions

$$Q = Ran_E Lan_{fR}(Q) = Ran_E Lan_R(Q)$$

and

$$fI = Lan_{fR} Ran_E(fI) = Lan_{fR} Ran_E(I)$$

are satisfied. Now let

$$Q = Re[Ran_E(ReJ)] \leqq {}_RE_{End({}_RE)}$$

and

$$I = Lan_R(Q) \leqq {}_RR_R.$$

Then since $Q \subseteq Ran_E(ReJ)$ we have

$$fI = Lan_{fR}(Q) \supseteq Lan_{fR} Ran_E(ReJ) = fReJ,$$

and since $Ie[Ran_E(ReJ)] \subseteq IQ = 0$ we also have

$$fIe \subseteq Lan_{fR} Ran_E(ReJ) = fReJ,$$

so that

$$fIe = fReJe.$$

Now since E is distributive and $Q \leq ReE$, we must have $Q/JQ \cong Re/Je$ by Lemma 15.13, so since Re is distributive, $Q \cong Re/Ke$ for some ideal $K \leqq {}_RR_R$. Thus since

$Q = \mathrm{Ran}_E \mathrm{Lan}_R(Q) = \mathrm{Ran}_E(I)$ we have

$$\mathrm{Ran}_E(I) \cong Re/Ke$$

and so according to Theorem 15.15

$$fR/fI \cong \mathrm{Lan}_{E(eR/eJ)}(K) \leqq E(eR/eJ).$$

Finally, applying Lemma 11.6 again we have an embedding of right modules

$$fRe/fReJe = fRe/fIe \hookrightarrow E(eR/eJ)e = E(eRe/eJe)$$

so that $fRe/fReJe$ is simple (or zero) over eRe as promised.

According to the classical Azumaya-Morita theorems (see Corollary 4.4 and Theorem 11.1) there is a duality between the categories of finitely generated left and right modules over a ring R if and only if R is artinian and the endomorphism ring of the minimal left cogenerator over R is isomorphic to the basic ring of R. The following result was proved by Belzner [90] and Fuller-Xue [91] independently. This theorem partially solves Azumaya's conjecture for locally distributive rings.

THEOREM 15.18. If R is a basic locally distributive ring and S is the endomorphism ring of the minimal left cogenerator over R, then S is locally distributive.

PROOF. Let $E = E_1 \oplus \ldots \oplus E_n$ with $E_i = E(Re_i/Je_i)$ and $i = 1,\ldots,n$. Since R is an exact artinian ring, there is a duality $D = \mathrm{Hom}(-, {}_RE_S)$ between the finitely generated left R-modules and right $S = \mathrm{End}({}_RE)$-modules; so for each finitely generated ${}_RM$ there is a lattice anti-isomorphism

$$\theta : L(M) \longrightarrow L(D(M))$$

via

$$\theta : N \longmapsto \mathrm{Ker}(D(i_N)), \qquad N \leq M$$

where i_N is the inclusion map $N \xrightarrow{i_N} M$. Letting $f_i \in S$

be the idempotent for E_i in the decomposition $E = E_1 \oplus \ldots \oplus E_n$, $f_1, \ldots, f_n$ is a basic set of idempotents for S. It follows that the indecomposable right projective S-module $f_iS \cong D(E_i)$ and right injective S-module $D(Re_i)$ are all distributive. Thus by Theorem 15.17, S is locally distributive.

The self-duality for locally distributive rings is still open. Belzner [90] provided a partial result whose proof is so technical that we only state the result as follows.

THEOREM 15.19. Let R be a locally distributive ring with a complete set of primitive idempotents $e_1, \ldots, e_n$.

(1) Let X be the set of pairs (i,j), $1 \leq i,j \leq n$, such that there exists some $k \in \{1,\ldots,n\}$ with $c(e_kRe_k) > 1$ and $e_iRe_kRe_j \neq 0$. Then $I_0 = \sum_{(i,j)\in X} \mathrm{Soc}(e_iRe_j)$ is an ideal of R.

(2) For any ideal I containing I_0 the factor ring R/I has a self-duality.

COROLLARY 15.20. Let R be as in Theorem 15.19. If each e_iRe_i is a division ring, then R has a self-duality.

PROOF. In this case $I_0 = 0$. Hence the conclusion follows from the theorem.

It is also interesting to give a second proof by showing that R is a finite normalizing extension of a division ring and then using Kraemer's Theorem 9.11 to obtain the self-duality of R. To show R has self-duality, we may assume that R is basic and indecomposable. If $e_iRe_j \neq 0$, then e_iRe_j is uniserial as left e_iRe_i- and right e_jRe_j-module, and there exists some $e_{ij} \in e_iRe_j$ such that $e_iRe_j = e_iRe_ie_{ij} = e_{ij}e_jRe_j$, hence here is an isomorphism of rings $e_iRe_i \cong e_jRe_j$ via $r_i \longmapsto r_j$ where $r_ie_{ij} = e_{ij}r_j$. Now R is indecomposable, so there is an isomorphism $f_i : e_1Re_1 \cong e_iRe_i$ for each $i = 2,\ldots,n$. Let $D = \{r_1+f_2(r_1)+\ldots+f_n(r_1) \mid r_1 \in e_1Re_1\} \cong e_1Re_1$ that is a division subring of R. Then $De_{ij} = e_{ij}D$ for each e_{ij} so $R = \Sigma_{i,j} De_{ij}$ is a normalizing extension over D.

SECTION 16. ARTINIAN DUO RINGS

A bimodule $_RE_S$ is called left (right) duo in case each R- (S-) submodule is also an S- (R-) submodule, and it is called duo if it is both left and right duo. This is a generalization of left duo (right duo, duo) rings. Using Lemma 14.5, one notes that artinian duo rings are exact. In this section, we prove that the endomorphism ring of the minimal injective cogenerator over an artinian duo ring is still an artinian duo ring, and that an artinian duo ring has self-duality if its Jacobson radical is a direct sum of the ideals with simple socles. These provide further evidence for Azumaya's conjecture that exact artnian rings have self-duality.

LEMMA 16.1. A bimodule $_RE_S$ is duo if and only if $Rx = xS$ for all $x \in E$.

PROOF. Straightfoward.

PROPOSITION 16.2. Let R be a left artinian ring with a duality induced by $_RE_S$ such that $c(_RE) = c(E_S)$. The following statements are equivalent:

(1) R is left duo;

(2) E is right duo;

(3) E is left duo;

(4) S is right duo.

PROOF. (2) ==> (3). Let $x \in E$, then $Rx \subseteq xS \leqq {_RE_S}$. Let $(Rx)^* = Hom_R(Rx,E)$ and $x(Rx)^* = \{(x)f \mid f \in (Rx)^*\}$. Since $_RE$ is injective, $x(Rx)^* \subseteq xS$. On the other hand, $xs = (x)(s|_{Rx}) \in x(Rx)^*$ for all $s \in S$. Hence $x(Rx)^* = xS$, and we have

$$c({}_R Rx) = c((Rx)^*_S) = c(x(Rx)^*_S) = c(xS_S) = c({}_R(xS)),$$

where the second equality holds since there is an isomorphism $(Rx)^*_S \cong x(Rx)^*_S$ given by $f \longmapsto (x)f$, and the last equality holds since ${}_R E_S$ is right duo, $xS \leqq {}_R E_S$ and $c({}_R E) = c(E_S)$. Thus since $Rx \subseteq xS$ it follows that $Rx = xS$.

(3) ==> (2). By a similar proof.

(1) <==> (2) and (3) <==> (4). By Theorem 2.6.

The next result is due to Courter [82, Theorem 2.2].

THEOREM 16.3. Let R be an artinian left duo ring. Then R is right duo if and only if $c({}_R R) = c(R_R)$.

PROOF. (==>). If R is an artinian duo ring, any composition series for ${}_R R$ is also a composition series for R_R.

(<==). Let R be an artinian left duo ring with $c({}_R R) = c(R_R)$. If I is a left ideal, then I is an ideal and it is easy to see that $c({}_R I) = c(I_R)$. Now we assume that R is not right duo, and let R be a counterexample with $n = c({}_R R)$ minimal. In particular each proper quotient of R is a duo ring and any right ideal containing a non-zero left ideal is an ideal. Let M be any maximal left ideal, then $D = R/M$ is a division ring. The right R-module $\mathrm{Ran}_R(M) \cong \mathrm{Hom}_R(R/M, {}_R R)$ is a left D-vector space and a left R-module. To show that $\mathrm{Ran}_R(M)$ is a minimal right ideal we shall show that it is a minimal left ideal.

Let $y \in R$ such that yR is contained properly in Ry: let $x \in \mathrm{Ran}_R(M)$. Since Dx is a left ideal, Dx and $yR + Dx$ are right ideals. Thus $yR + Dx$ is an ideal and contains Ry. We have

$$yR + Dx = Ry + Dx.$$

As right modules, $c(Ry) \leqq c(yR + Dx) \leqq c(yR) + 1 \leqq c(Ry)$, so that

$$Ry = yR + Dx, \qquad yR \cap Dx = 0.$$

Suppose $\dim({}_D\mathrm{Ran}_R(M)) > 1$, say $Dx \oplus Dx_1 \leq \mathrm{Ran}_R(M)$. Then

$$Ry = yR + Dx = yR + Dx_1 = yR + (Dx \oplus Dx_1).$$

As right modules, $c(yR + (Dx \oplus Dx_1)) < c(yR) + 2$, so that $yR \cap (Dx \oplus Dx_1) \neq 0$. For some non-zero $x_2 \in \mathrm{Ran}_R(M)$ the ideal Dx_2 is contained in yR making yR an ideal. This contradiction proves that $\mathrm{Ran}_R(M)$ is one dimensional and is a minimal left ideal. Hence $\mathrm{Ran}_R(M)$ is a minimal right ideal. Since the dual of simple left R-modules are simple and $c({}_RR) = c(R_R)$, R is a QF-ring by Theorem 11.16. It follows that R has the double annihilator property, so each right ideal is a right annihilator of an ideal, whence it is an ideal. The ring R is right duo.

Each idempotent in a duo ring is central (Lemma 12.2), so an artinian duo ring is a finite product of local artinian duo rings. Let R be a local artinian duo ring and $J = J(R)$. Habeb [89] proved that if R/J is a non-commutative division ring then R is a uniserial ring, and that if $J^2 = 0$ and R/J is a field then $S = \mathrm{End}({}_RE(R/J))$ is a local artinian duo ring with $J(S)^2 = 0$. Now we prove Habeb's conjecture that S is always an artinian duo ring without assuming that $J^2 = 0$. The rest results in this section are due to Xue [89].

THEOREM 16.4. If R is an artinian duo ring with the minimal injective cogenerator ${}_RE$ and $S = \mathrm{End}({}_RE)$, then ${}_RE_S$ is duo and S is an artinian duo ring.

PROOF. The hypothesis implies that the exact bimodule ${}_RE_S$ defines a duality and that R and S are basic exact artinian rings. Then $c({}_RE) = c(E_S)$ and $c({}_SS) = c(S_S)$ by the results in Section 14 (see the discussion preceding Proposition 14.6). Now the results follow from Proposition 16.2 and Theorem 16.3.

A module ${}_RM$ is called (<u>local</u>) <u>colocal</u> in case $M/\mathrm{Rad}({}_RM)$

$(\mathrm{Soc}({}_R M))$ is simple. Now we show that there is a large class of artinian duo rings which do have self-duality. In particular, we prove Azumaya's conjecture for duo rings with radical square zero.

THEOREM 16.5. If R is an artinian duo ring such that $J(R)$ is a direct sum of colocal ideals, then R has self-duality.

PROOF. Assume that R is local. Let ${}_R E = E(R/J(R))$ and $S = \mathrm{End}({}_R E)$. Then ${}_R E_S$ is duo and S is a local artinian duo ring by the above theorem. We need to show $R \cong S$.

Since $J(R)$ is a direct sum of colocal ideals, it follows that $E_S/\mathrm{Soc}(E_S)$ is a direct sum of local submodules. So there are $x_1, \ldots, x_n \in E$ such that

$$E/\mathrm{Soc}(E_S) = \oplus_{i=1}^n \bar{x}_i S = \oplus_{i=1}^n R\bar{x}_i.$$

And then

$$E = \Sigma_{i=1}^n x_i S = \Sigma_{i=1}^n Rx_i.$$

from the fact that R local implies $\mathrm{Soc}(E_S) \leq x_i S$ for all i. We may assume that $n \geqq 2$ (If $n = 1$, R is QF). Using results in Section 2 we have

$$\begin{aligned} &\mathrm{Ran}_E\Big[\mathrm{Lan}_R(x_i) + (\cap_{j\neq i} \mathrm{Lan}_R(x_j))\Big] \\ &= \mathrm{Ran}_E\mathrm{Lan}_R(x_i) \cap \Big[\Sigma_{j\neq i} \mathrm{Ran}_E\mathrm{Lan}_R(x_j)\Big] \\ &= x_i S \cap (\Sigma_{j\neq i} x_j S) = \mathrm{Soc}(E_S) = \mathrm{Soc}({}_R E) = \mathrm{Ran}_E(J(R)), \end{aligned}$$

so

$$J(R) = \mathrm{Lan}_R(x_i) + (\cap_{j\neq i} \mathrm{Lan}_R(x_j)), \qquad i = 1,\ldots,n.$$

Similarly, we have

$$J(S) = \mathrm{Ran}_S(x_i) + (\cap_{j\neq i} \mathrm{Ran}_S(x_j)), \qquad i = 1,\ldots,n.$$

Now since $Rx_i = x_i S$, we have a ring isomorphism

$$\phi_i : R/\mathrm{Lan}_R(x_i) \cong S/\mathrm{Ran}_S(x_i)$$

given by

$$\phi_i : \bar{r} \longmapsto \bar{s} \qquad \text{if} \quad rx_i = x_i s.$$

So we get a ring isomorphism

$$\phi = \oplus_i \phi_i : \oplus_i R/\mathrm{Lan}_R(x_i) \cong \oplus_i S/\mathrm{Ran}_S(x_i).$$

And since

$$\cap_i \mathrm{Lan}_R(x_i) = \mathrm{Lan}_R(\Sigma_i x_i S) = \mathrm{Lan}_R(E) = 0,$$

we have a canonical embedding of rings

$$f : R \hookrightarrow \oplus_i R/\mathrm{Lan}_R(x_i).$$

Similarly, we have a canonical embedding of rings

$$g : S \hookrightarrow \oplus_i S/\mathrm{Ran}_S(x_i).$$

To show $R \cong S$, we only need to show that

$$\phi(\mathrm{Im}(f)) = \mathrm{Im}(g).$$

We shall show $\phi(\mathrm{Im}(f)) \subseteq \mathrm{Im}(g)$, since the other containment can be proved similarly. To do so, we show that for each $r \in R$ there is an $s \in S$, such that

$$rx_i = x_i s, \qquad \text{for all } i.$$

Let $r \in R$, and we define an $s \in S = \mathrm{End}({}_R E)$ via

$$s : \Sigma_i r_i x_i \longmapsto \Sigma_i r_i r x_i, \qquad \text{for } \Sigma_i r_i x_i \in {}_R E = \Sigma_i Rx_i.$$

If $\Sigma_i r_i x_i = 0$, each $r_i \in J(R) = \mathrm{Lan}_R(x_i) + \cap_{j \neq i} (\mathrm{Lan}_R(x_j))$, and hence $r_i = y_i + z_i$ for some $y_i \in \mathrm{Lan}_R(x_i)$, and $z_i \in \cap_{j \neq i} (\mathrm{Lan}_R(x_j))$. Since each $\mathrm{Lan}_R(x_i)$ is an ideal of R, we have

$$\begin{aligned} \Sigma_i r_i r x_i &= \Sigma_i (y_i + z_i) r x_i = \Sigma_i z_i r x_i = (\Sigma_j z_j) r (\Sigma_i x_i) \\ &= (\Sigma_j z_j)(\Sigma_i x_i) s' = (\Sigma_i z_i x_i) s' = (\Sigma_i y_i + z_i) x_i) s' \\ &= (\Sigma_i r_i x_i) s' = 0, \end{aligned}$$

where $r(\Sigma_i x_i) = (\Sigma_i x_i) s'$ for some $s' \in S$ since ${}_R E_S$ is duo. Hence s is well-defined, and then $s \in S$. By the definition of s, we have $rx_i = x_i s$ for all i.

COROLLARY 16.6. If R is an artinian duo ring with $J(R)^2 = 0$, then R has self-duality.

Now we give an artinian duo ring that satisfies the condition of Theorem 16.5. Mano [84] gave a local uniserial ring as follows:

Let F be a field and let $K = F(x)$ be the field of rational functions over F with formal derivative $(\ )'$. Put

$$A = K \times K \times K$$

as abelian group. For $(k_1, k_2, k_3), (l_1, l_2, l_3) \in A$, define $(k_1, k_2, k_3)(l_1, l_2, l_3) = (k_1l_1, k_1l_2 + k_2l_1, k_1l_3 + k_2l_2 + k_3l_1 + k_1'l_2)$. Then one checks that A is a non-commutative local uniserial ring of length 3 such that $A/J(A)^2$ is commutative.

EXAMPLE 16.7. Let A be a local uniserial ring of length $n \geqq 2$ with $A/J(A)^{n-1}$ commutative (e.g., the ring A given above). Let $J = J(A)$ and $R = A \propto (J/J^2)$, the trivial extension of A by J/J^2. Then R is a local artinian duo ring: Since A is a local uniserial ring, $J = Ap = pA$ for some $p \in J\backslash J^2$ and then $0 \neq \bar{p} = p + J^2 \in J/J^2$. Let $(a, b\bar{p}) \in R$. If $a \in A\backslash J$, then $(a, b\bar{p})$ is a unit of R and $R(a, b\bar{p}) = R = (a, b\bar{p})R$; so let $(ap, b\bar{p}) \in R$, then for $(c, d\bar{p}) \in R$, we have

$$\begin{aligned}(ap, b\bar{p})(c, d\bar{p}) &= (apc, b\bar{p}c)\\ &= (ac_1p, bc_1\bar{p})\\ &= (c_1ap, c_1b\bar{p}) \quad (\text{since } A/J^{n-1} \text{ is commutative})\\ &= (c_1, 0)(ap, b\bar{p}) \in R(ap, b\bar{p}),\end{aligned}$$

where $pc = c_1p$ for some $c_1 \in A$ since $Ap = pA$. So we have $(ap, b\bar{p})R \subseteq R(ap, b\bar{p})$. The other containment can be proved similarly, and so R is duo. Now

$$J(R) = J \times (J/J^2) = (J \times 0) \oplus (0 \times J/J^2),$$

where $J \times 0$ is a uniserial ideal and $0 \times J/J^2$ is a simple ideal of R. Hence this ring R has self-duality by Theorem 16.5. If A is not commutative, neither is R since $R/(0 \times J/J^2) \cong A$ as rings.

Dischinger and Müller [84] and Waschbüsch [86] proved that serial rings have self-duality. The above example is not cover by this theorem, since it is not serial. However this example is a trivial extension of a serial ring.

CHAPTER 5
OTHER TYPES OF RINGS WITH DUALITY

In this final chapter, duality for other types of rings are considered. In Section 17, we study left noetherian rings with a duality. Most results are due to Müller [69], Jategaonkar [81], and Menini [86]. If R is a left or right perfect ring then the duality of R implies that R is left artinian. This is done in Section 18, the final section of the book. As a generalization of perfect rings, Camillo and Xue [91] introduced quasi-perfect rings. The behavior of such rings with dulaity are presented, e.g., if ${}_RE_S$ defines a duality then R is left quasi-perfect if and only if S is right quasi-perfect.

SECTION 17. NOETHERIAN RINGS

In this section, we consider noetherian rings with a duality. Muller [69] proved that if R has a duality induced by ${}_RE_S$ and $\cap_{n=1}^{\infty} J(R)^n = 0$ then R is left noetherian, S is right noetherian, and both ${}_RE$ and E_S are artinian modules. Menini's example [86] shows that the condition that $\cap_{n=1}^{\infty} J(R)^n = 0$ is essential.

LEMMA 17.1. Let R be a ring such that $\cap_{n=1}^{\infty} J(R)^n = 0$. If R is left linearly compact then R is left noetherian.

PROOF. Let $J = J(R)$ and I a left ideal. We want to show that I is finitely generated. Since ${}_RR$ is linearly compact, I/JI is a linearly compact semisimple module, so I/JI is finitely generated. Let $I = \Sigma_{i=1}^{m} Ra_i + JI$. Then $J^kI = \Sigma_{i=1}^{m} J^ka_i + J^{k+1}I \subseteq \Sigma_{i=1}^{m} Ra_i + J^{k+1}I$, and so

$I = \Sigma_{i=1}^{m} Ra_i + J^k I$ for each $k = 1,2,\ldots.$ Since $\cap_{k=1}^{\infty} J^k I \subseteq \cap_{k=1}^{\infty} J^k = 0$ so using Corollary 3.9, we get $I = \cap_{k=1}^{\infty}(\Sigma_{i=1}^{m} Ra_i + J^k I) = \Sigma_{i=1}^{m} Ra_i + \cap_{k=1}^{\infty} J^k I = \Sigma_{i=1}^{m} Ra_i$.

THEOREM 17.2 (Müller). Let R be a ring such that $\cap_{n=1}^{\infty} J(R)^n = 0$. If R has a duality induced by ${}_RE_S$ then R is left noetherian, S is right noetherian, both ${}_RE$ and E_S are artinian modules, and $E = \cup_{n=1}^{\infty} \mathrm{Ran}_E(J(R)^n) = \cup_{n=1}^{\infty} \mathrm{Lan}_E(J(S)^n)$.

PROOF. Since ${}_RE_S$ defines a duality, both ${}_RR$ and S_S are linearly compact. Since $\cap_{n=1}^{\infty} J(R)^n = 0$, by Theorem 2.6 and Lemma 14.4 we have

$$\begin{aligned} E &= \mathrm{Ran}_E(\cap_{n=1}^{\infty} J(R)^n) = \Sigma_{n=1}^{\infty} \mathrm{Ran}_E(J(R)^n) \\ &= \Sigma_{n=1}^{\infty} \mathrm{Lan}_E(J(S)^n) = \mathrm{Lan}_E(\cap_{n=1}^{\infty} J(S)^n), \end{aligned}$$

and then

$$\cap_{n=1}^{\infty} J(S)^n = \mathrm{Ran}_S(E) = 0.$$

By Lemma 17.1, R is left noetherian and S is right noetherian. Then ${}_RE$ and E_S are artinian modules by Theorem 2.6.

Recall that if R is a left artinian ring with a duality induced by ${}_RE_S$ then S is right artinian. Now we take an example of Menini [86] to show that an analogous statement for left noetherian rings is not true.

Let F be a field, then $F[[x]]$ is a local commutative noetherian domain with a self-duality induced by $F[1/x]$ (see Example 10.9). Let $F((x))$ be the quotient field of $F[[x]]$, then $F((x))$ is a linearly compact $F[[x]]$-module by Lemma 10.11.

EXAMPLE 17.3. Let $R = \begin{bmatrix} F((x)) & F((x)) \\ 0 & F[[x]] \end{bmatrix}$, which is a left noetherian ring. Since ${}_{F[[x]]}F((x))$ is linearly compact (Lemma 10.11), by Corollary 10.3, the left noetherian ring R has a duality induced by ${}_RE = \begin{bmatrix} F((x)) & 0 \\ F((x)) & F[1/x] \end{bmatrix}$. One checks that ${}_RE$ is not artinian since ${}_{F[[x]]}F((x))$ is not artinian. It follows that $S = \mathrm{End}({}_RE)$ is not right noetherian. In this example, one sees that $J(R) = \begin{bmatrix} 0 & F((x)) \\ 0 & xF[[x]] \end{bmatrix}$, and then $\cap_{n=1}^{\infty} J(R)^n = \begin{bmatrix} 0 & F((x)) \\ 0 & 0 \end{bmatrix} \neq 0$.

Jacobson [64] asked a question that gave rise to the so-call Jacobson's Conjecture:

"If R is a left noetherian ring with Jacobson radical J, then $\cap_{n=1}^{\infty} J^n = 0$".

Herstein [65] and Jategaonkar [68] have constructed counterexamples to show that this conjecture is not true. Menini's Example 17.3 shows that this conjecture is not true even for left noetherian rings with a duality. This settles Jategarnkar's Problem 1 [81] in the negative. It should be noted that Jacobson's Conjecture is still open for (two-sided) noetherian rings. Menini [86] proved that the conjecture is true for noetherian left linearly compact rings.

LEMMA 17.4. Let R be a ring with $J = J(R)$ and $J^{\infty} = \cap_{n=1}^{\infty} J^n$. Suppose that R is left linearly compact and ${}_RJ$ is finitely generated. Then $J^{\infty} = J^{\infty}J$.

PROOF. Let $J = \Sigma_{i=1}^{n} Ra_i$. Define an R-homomorphism

$$f:\ {}_RR^{(n)} \longrightarrow {}_RJ$$

by setting

$$(r_1, \ldots, r_n) \longmapsto r_1a_1 + \ldots + r_na_n.$$

Now we have

$$f(\cap_{k=1}^{\infty}(J^kR^{(n)}) = f((J^{\infty})^{(n)}) = J^{\infty}J.$$

As ${}_RR^{(n)}$ is linearly compact, by Theorem 3.8 we have

$f(\cap_{k=1}^{\infty}(J^k R^{(n)}) = \cap_{k=1}^{\infty} f(J^k R^{(n)}) = \cap_{k=1}^{\infty} J^k = J^{\infty}$.

COROLLARY 17.5. Let R be a ring with $J = J(R)$ and $J^{\infty} = \cap_{n=1}^{\infty} J^n$. Suppose that R is left linearly compact and that both ${}_R J$ and $(J^{\infty})_R$ are finitely generated. Then $J^{\infty} = 0$.

PROOF. By Lemma 17.4, $J^{\infty} = J^{\infty} J$. Apply now Nakayama's Lemma.

REMARK 17.6. Corollary 17.5 holds if R is a noetherian left linearly compact ring. In particular, if R is a noetherian ring with a duality, then $\cap_{n=1}^{\infty} J(R)^n = 0$. This was first proved, in another way, by Jategaonkar [81b].

The following question was raised by Jategaonkar [81], and it is still open now. The analogous question for artinian rings is also open. (See Question 11.18.)

QUESTION 17.7. Let R be a (two-sided) noetherian ring with a duality induced by ${}_R E_S$. According to Remark 17.6, $\cap_{n=1}^{\infty} J(R)^n = 0$. Hence by Theorem 17.2, S is right noetherian. Is S necessarily left noetherian?

SECTION 18. PERFECT AND QUASI-PERFECT RINGS

A ring R is called <u>left</u> (<u>right</u>) <u>perfect</u> in case each of its left (right) modules has a projective cover. As a generalization of perfect rings, Camillo and Xue [91] called a ring R <u>left</u> (<u>right</u>) <u>quasi-perfect</u> in case each left (right) artinian R-module has a projective cover, and proved that the class of left quasi-perfect rings lies strictly between that of left perfect rings and that of semiperfect rings. A left perfect ring need not be right perfect (Anderson-Fuller [76, p.322]); in fact, this left perfect ring is not right quasi-perfect (see Miller-Turnidge [73]). In this section we consider left (quasi-) perfect rings with a duality. It turns out that a left or right perfect ring with a duality is left artinian (Corollary 18.4). Then we turn our attention to quasi-perfect rings with a duality. We prove that if R is a left quasi-perfect ring with a duality ${}_RE_S$ then S is right quasi-perfect.

The pioneering work on perfect rings was done by Bass [60] and the following principal characterizations of left perfect rings are contained in Bass [60, Theorem P]. A subset I of a ring R is called left (right) T-nilpotent in case for every sequence $a_1, a_2, \ldots$ in I there is an n such that $a_1a_2\ldots a_n = 0$ ($a_n\ldots a_2a_1 = 0$). One sees that a left ideal J is left T-nilpotent if and only if JM is superfluous in M for every non-zero left R-module M. (For proofs, see Anderson-Fuller [74].)

THEOREM 18.1 (Bass [60]). Let R be a ring with radical J = J(R). Then the following statements are equivalent:

(1) R is left perfect;

(2) R is semilocal and J is left T-nilpotent;

(3) R is semilocal and every non-zero left R-module contains a maximal submodule;

(4) Every flat left R-module is projective;

(5) R satisfies the minimum condition for principal right ideals;

(6) R contains no infinite orthogonal set of idempotents and every non-zero right R-module contains a minimal submodule.

It is immediate that each semiprimary ring is both left and right perfect. The following two results are contained in Sandomierski [72b].

PROPOSITION 18.2. Let R be a left perfect ring. If ${}_RM$ is a linearly compact module, then ${}_RM$ is noetherian.

PROOF. Let N be a submodule of M. Since N/J(R)N is semisimple and linearly compact, it is finitely generated. Let

$N = \Sigma_{i=1}^{n} Rm_i + J(R)N$, then $N = \Sigma_{i=1}^{n} Rm_i$ is finitely generated since $J(R)N$ is superfluous in N.

PROPOSITION 18.3. Let R be a right perfect ring. If ${}_RM$ is a linearly compact module, then ${}_RM$ is artinian.

PROOF. We show that each quotient module M/N is finitely cogenerated. Since M/N is linearly compact, it has a finitely generated socle $Soc(M/N)$ that is essential in M/N by Theorem 18.1(6).

It follows from the above two propositions that a linearly compact module over a (left and right) perfect ring has finite length. This is a generalization of Corollary 3.5.

COROLLARY 18.4. Let R be a left linearly compact ring (in particular, if R has a duality). If R is left or right perfect then it is left artinian.

PROOF. If R is left linearly compact and right perfect, then R is left artinian by Proposition 18.3.

Now let R be left linearly compact and left perfect. By Proposition 18.2, R is left noetherian. Levitzki's Theorem (see Anderson and Fuller [74, Theorem 15.22]) states that each one-sided ideal in a left noetherian ring is nilpotent. Since $J(R)$ is left T-nilpotent, it is nil and hence it is nilpotent by Levitzki's Theorem. It follows that R is semiprimary, and so R is left artinian by Hopkin's Theorem (see Section 1).

Now we turn our attention to quasi-perfect rings. Next we characterize left quasi-perfect rings via their left artinian modules.

THEOREM 18.5. The following are equivalent for a semiperfect ring R:

(1) R is left quasi-perfect;

(2) Every non-zero left artinian R-module has a maximal submodule;

(3) Every left artinian R-module has finite length;

(4) Every left artinian R-module is finitely generated.

PROOF. (1) ==> (2). If ${}_RM \neq 0$ is artinian, then there is a projective cover ${}_RP$ with superfluous submodule K such that $M \cong P/K$. It is well-known (or see Anderson and Fuller [74]) that projective module P has a maximum submodule L, then $K \subseteq L$ since K is superfluous in P. Thus L/K is a maximum submodule in $P/K \cong M$.

(2) ==> (3). If ${}_RM \neq 0$ is artinian, then $M = M_0$ has a maximum submodule M_1 that is still artinian. If $M_1 \neq 0$, M_1 has a maximum submodule M_2, ... Since M is artinian, M_n is simple for some $n \geqq 0$. Thus

$$M = M_0 > M_1 > \ldots > M_n > 0$$

is a composition series of M.

(3) ==> (4). Clearly.

(4) ==> (1). A ring is semiperfect if and only if each finitely generated module has a projective cover (Theorem 1.18).

Using the above characterization we can easily show that left quasi-perfect rings are preserved under Morita equivalence.

THEOREM 18.6. If R and S are two Morita equivalent rings, then R is left quasi-perfect if and only if S is left quasi-perfect.

PROOF. Let $F : R \approx S$ define an equivalence and S a left quasi-perfect ring. Since the property of semiperfect rings is Morita invariant, R is semiperfect. If ${}_RM$ is an R-module, then ${}_RM$ is artinian (has finite length) if and only if ${}_SFM$ is artinian (has finite length). Thus by Theorem 18.5, R is left quasi-perfect.

Suppose that R is a semiperfect ring that is not left quasi-perfect. Then by Theorem 18.5 there exists a non-zero left artinian R-module ${}_RA$ without maximum submodule. So $JA = A$. Let $\Sigma = \{ 0 \neq B \leq {}_RA \mid JB = B \}$. Since ${}_RA$ is artinian, Σ has a minimal element B with the property: (1) $JB = B$; and (2) ${}_RB$ has infinite length but any proper submodule of ${}_RB$ has finite length.

PROPOSITION 18.7. Let R be a commutative semiperfect ring. If J is nil then R is quasi-perfect.

PROOF. If R is not quasi-perfect, then by the above remark there is a non-zero artinian R-module B with $JB = B$ and any proper submodule of B has finite length. For any $j \in J$, j is nilpotent, so $jB \neq B$. Thus jB is a proper submodule of B, and hence jB has finite length. Now $J(jB) = j(JB) = jB$, so $jB = 0$ by Nakayama's Lemma. It follows that $B = JB = 0$, a contradiction.

The following two examples show that a commutative quasi-perfect ring need not be perfect, and a commutative local (hence semiperfect) ring need not be left quasi-perfect.

EXAMPLE 18.8. Let $GF(p)$ be the field with p elements and $GF(p)[x_1,x_2, \ldots]$ the polynomial ring with commutative indeterminates. Let $R = GF(p)[x_1,x_2, \ldots]/\langle x_1^p,x_2^p,\ldots\rangle$. Then R is a local commutative ring with radical $J = \langle \bar{x}_1,\bar{x}_2, \ldots\rangle$, and $j^p = 0$ for all $j \in J$. By Proposition 18.7, R is quasi-perfect. But J is not T-nilpotent, so R is not perfect.

EXAMPLE 18.9. Let F be a field, then $R = F[[x]]$ is a commutative local ring and its minimal injective cogenerator ${}_RE = F[1/x]$ is artinian but not noetherian. Hence R is not quasi-perfect by Theorem 18.5.

To see that quasi-perfect-ness is also preserved by Morita duality, we make the following observations. It is well-known (see Anderson-Fuller [74]) that the endomorphism ring of a module with finite length is semiprimary.

LEMMA 18.10. Let ${}_RE_S$ define a duality. If R is left quasi-perfect and left noetherian, then R is left artinian.

Proof. If $\cap_{n=1}^{\infty} J(R)^n = 0$, then by Theorem 17.2. ${}_RE$ is artinian and S is right noetherian, hence ${}_RE$ has finite length (Theorem 18.5). Then $S = \mathrm{End}({}_RE)$ is semiprimary, so it is right artinian. Therefore R is left artinian.

If $\cap_{n=1}^{\infty} J(R)^n \neq 0$, let $\bar{R} = R/\cap_{n=1}^{\infty} J(R)^n$. It is easy to see that $\bar{R}$ is left quasi-perfect. Now $\bar{R}$ has a duality by Corollary 2.5. Hence from the above result, $\bar{R}$ is left artinian. Then $J(R)^n = J(R)^{n+1}$ for some n. By Nakayama Lemma's $J(R)^n = 0$. This contradicts to that $\cap_{n=1}^{\infty} J(R)^n \neq 0$.

THEOREM 18.11. Let ${}_RE_S$ define a duality. The following are equivalent:

(1) R is left quasi-perfect;

(2) For each ideal I in R, if R/I is left noetherian then R/I is left artinian;

(3) For each ideal I in R, if $\cap_{n=1}^{\infty} J(R/I)^n = 0$ then R/I is left artinian;

(4) For each ideal K in S, if $\cap_{n=1}^{\infty} J(S/K)^n = 0$ then S/K is right artinian;

(5 For each ideal K in S, if S/K is right noetherian then S/K is right artinian;

(6) S is right quasi-perfect.

PROOF. (1) ==> (2). Since R/I is still left quasi-perfect, the result follows from Corollary 2.5 and Lemma 18.10.

(2) ==> (3). By Corollary 2.5 and Theorem 17.2, R/I is left noetherian.

(3) ==> (1). Suppose R is not left quasi-perfect, then there is an artinian R-module ${}_RB$ with infinite length such that any proper submodule of ${}_RB$ has finite length. Thus $\bar{R} = R/\mathrm{Lan}_R(B)$ is still not a left quasi-perfect ring by Theorem 18.5, since ${}_{\bar{R}}B$ is an artinian $\bar{R}$-module with infinite length. For each $b \in B$, Rb has finite length so $\bar{J}^n b = 0$ for some $n \geqq 1$, where $\bar{J} = J(\bar{R})$ is the radical of $\bar{R}$. Thus $(\cap_{n=1}^{\infty}\bar{J}^n)B = 0$. But B is a faithful $\bar{R}$-module, so $\cap_{n=1}^{\infty}\bar{J}^n = 0$. By Theorem 17.2, $\bar{R}$ is left noetherian since $\bar{R}$ still has a duality by Corollary 2.5. Now $\bar{R}$ is not left artinian since it is not left quasi-perfect. The contradiction proves the implication of (3) ==> (1).

Similarly, we have (4) <==> (5) <==> (6).

(4) ==> (3). Let I be an ideal of R and $\cap_{n=1}^{\infty}J(R/I)^n = 0$. By Corollary 2.5, $\mathrm{Ran}_E(I)$ defines a duality between left R/I-modules and right $\bar{S} = S/\mathrm{Ran}_S\mathrm{Ran}_E(I)$-modules. By Theorem 17.2, $\cap_{n=1}^{\infty}J(\bar{S})^n = 0$. Hence $\bar{S}$ is right artinian by (4). Then R/I is left artinian.

Similarly, we have (3) $\Rightarrow$ (4).

In Theorem 18.5 we characterize left quasi-perfect rings via their left artinian modules. Our concluding theorem characterizes these rings with duality via their left noetherian modules.

THEOREM 18.12. Let ${}_RE_S$ define a Morita duality. The following are equivalent:

(1) R is left quasi-perfect;

(2) Every left noetherian R-module has finite length;

(3) Every non-zero left noetherian R-module has a minimal submodule;

(4) Every left noetherian R-module is finitely cogenerated.

PROOF. (1) ==> (2). By Theorem 18.11, S is right quasi-perfect. If ${}_RM$ is noetherian then $\mathrm{Hom}_R(M,E)_S$ is artinin. So $\mathrm{Hom}_R(M,E)_S$ has finite length, and then ${}_RM$ has finite length.

(2) ==> (3) and (3) ==> (4). Obvious.

(4) ==> (1). Let N_S be artinian. Then ${}_R\mathrm{Hom}_S(N,E)$ is

noetherian and hence finitely cogenerated. Thus N_S is finitely generated. By Theorem 18.5, S is right quasi-perfect. So R is left quasi-perfect by THeorem 18.11.

BIBLIOGRAPHY

[68] I.K. Amdal and F. Ringdal, Catégories unisérielles, C.R. Acad. Sci. Paris, Serie A, 267 (1968), 85-87, 247-249.

[74] F.W. Anderson and K.R. Fuller, Rings and Categories of Modules, Springer-Verlag, Berlin-New York, 1974.

[82] P.N. Ánh, Duality of modules over topological rings, J. Algebra 75 (1982), 395-425.

[90] P.N. Ánh, Morita duality for commutative rings, Comm. Algebra 18 (1990), 1781-1788.

[91] P.N. Ánh, Characterisation of two-sided PF-rings, J. Algebra 141 (1991), 316-320.

[59] G. Azumaya, A duality theory for injective modules, Amer. J. Math. 81 (1959), 249-278.

[66] G. Azumaya, Completely faithful modules and self-injective rings, Nagoya Math. J. 27 (1966), 697-708.

[83] G. Azumaya, Exact and serial rings, J. Algebra 85 (1983), 477-489.

[60] H. Bass, Finitistic dimension and a homological generalization of semiprimary rings, Trans. Amer. Math. Soc. 95 (1960), 466-488.

[90] Thomas Belzner, Towards self-duality of semidistributive artinian rings, J. Algebra 135 (1990), 74-95.

[75] V.P. Camillo, Distributive modules, J. Algebra 36 (1975), 16-25.

[86] V.P. Camillo, K.R. Fuller and J.K. Haack, On Azumaya's exact rings, Math. J. Okayama Univ. 28 (1986), 41-51.

[91] V.P. Camillo and Weimin Xue, On quasi-perfect rings, Comm. Algebra 19 (1991), 2841-2850.

[66] P.M. Cohn, On a class of binomial extensions, Illinois J. Math. 10 (1966), 418-424.

[84] R.R. Colby and K.R. Fuller, QF-3' rings and Morita duality, Tsukuba J. Math. 8 (1984), 183-188.

[82] R.C. Courter, Finite dimensional right duo algebras are duo, Proc. Amer. Math. Soc. 84 (1982), 157-161.

[58] J. Dieudonné, Remarks on quasi-Frobenius rings, Illinois J. Math. 2 (1958), 346-354.

[84] D. Dikranjan and A. Orsatti, On the structure of linearly compact rings and their dualities, Rend. Accad. Naz. Sci. XL Mem. Mat., 102, Vol.VIII, Fasc.9 (1984), 143-184.

[84] F. Dischinger and W. Müller, Einreihig zerlegbare artinsch ringe sind selbstdual, Arch. Math. 43 (1984), 132-136.

[86] F. Dischinger and W. Müller, Left PF is not right PF, Comm. Algebra 14 (1986), 1223-1227.

[72] V. Dlab and C.M. Ringel, Balanced rings, Lectures Notes in Mathematics 246, pp.73-142, Springer-Verlag, Berlin-New York, 1972.

[80] P. Dowbor, C.M. Ringel and D. Simson, Hereditary artinian rings of finite representation type, Lecture Notes in Mathematics 832, Springer-Verlag, Berlin-New York, 1980.

[76] C. Faith, Algebra II Ring Theory, Springer-Verlag, Berlin-New York, 1976.

[79] C. Faith, Self-injective rings, Proc. Amer. Math. Soc. 77 (1979), 157-164.

[75] R.M. Fossum, P.A. Griffith and I. Reiten, Trivial extensions of abelian categories, Lecture Notes in Math. 456, Springer-Verlag, Berlin and New York, 1975.

[69] K.R. Fuller, On indecomposable injectives over artinian rings, Pacific J. Math. 29 (1969), 115-135.

[70] K.R. Fuller, Double centralizers of injectives and projectives over Artinian rings, Illinois J. Math. 14 (1970), 658-664.

[78] K.R. Fuller, Rings of left invariant module type, Comm. Algebra 6 (1978), 153-167.

[F] K.R. Fuller, Problems relating representation and duality theories, unpublished.

[F2] K.R. Fuller, Unpublished Lecture Notes.

[89] K.R. Fuller, Artinian Rings (a supplement to Rings and Categories of Modules), Notas de Matematica Vol.2, Universidad de Murcia, 1989.

[81] K.R. Fuller and J.K. Haack, Duality for semigroup rings, J. Pure Appl. Algebra 22 (1981), 113-119.

[91] K.R. Fuller and Weimin Xue, On distributive modules and locally distributive rings, Chinese Ann. Math. Ser.B 12 (1991), 26-32.

[91] J.L. Gómez Pardo and N. Rodriguez Gonzalez, Endomorphism rings with Morita duality, Comm. Algebra 19 (1991), 2097-2112.

[76] K.R. Goodearl, Ring Theory: Nonsingular Rings and Modules, Marcel Dekker, New York, 1976.

[79] J.K. Haack, Self-duality and serial rings, J. Algebra 59 (1979), 345-363.

[82] J.K. Haack, Finite subdirect products of rings and Morita duality, Comm. Algebra 10 (1982), 2107-2119.

[89] J.M. Habeb, On Azumaya's exact rings and artinian duo rings, Comm. Algebra 17 (1989), 237-245.

[65] I.N. Herstein, A counterexample in noetherian ring, Proc. Nat. Acad. Sci. U.S.A. 54 (1965), 1036-1037.

[64] N. Jacobson, Structure of Rings, Amer. Math. Soc. Colloquium Publ. Vol. 37, Providence, RI, 1964 (revised edition).

[68] A.V. Jategaonkar, Left principal ideal domains, J. Algebra 8 (1968), 148-155.

[81] A.V. Jategaonkar, Morita duality and Noetherian rings, J. Algebra 69 (1981), 358-371.

[81b] A.V. Jategaonkar, Noetherian bimodules, primary decomposition and Jacobson's conjecture, J. Algebra 71 (1981), 379-400.

[61] R.E. Johnson and E.T. Wong, Quasi-injective modules and irreducible rings, J. London Math. Soc. 36 (1961), 260-268.

[82] F. Kasch (translated by D.A.W. Wallace), Modules and Rings, Academic Press, London-New York, 1982.

[68] T. Kato, Some generalizations of QF-rings, Proc. Japan Acad. 44 (1968), 114-119.

[81] Y. Kitamura, Quasi-Frobenius extensions with Morita duality, J. Algebra 73 (1981), 275-286.

[87] J. Kraemer, Charaterizations of the existence of (quasi-) self-duality for complete tensor rings, Algebra Berichte 56, Verlag Reinhard Fischer, Munchen 1987.

[87b] J. Kraemer, Self-duality for rings related to skew polynomials, J. Algebra 106 (1987), 490-509.

[88] J. Kraemer, (Self-) duality and the Picard group, Comm. Algebra 16 (1988), 2283-2311.

[90] J. Kraemer, Self-duality for finite normalizing extensions of skew fields, Math. J. Okayama Univ. 32 (1990), 103-109.

[84] B. Lemonnier, Dimension de Krull et dualite de Morita dans les extensions triangulaires, Comm. Algebra 12 (1984), 3071-3110.

[86] B. Lemonnier, Dimension et codimension de Gabriel dans les extensions triangulaires, Comm. Algebra 14 (1986), 941-950.

[55] H. Leptin, Linear kompakte moduln und ringe, Math. Z. 63 (1955), 241-267.

[84] T. Mano, Self-duality and ring extensions, J. Pure Appl. Algebra 32 (1984), 51-57.

[84b] T. Mano, Uniserial rings and skew polynomial rings, Tokyo J. Math. 7 (1984), 209-213.

[58] E. Matlis, Injective modules over noetherian rings, Pacific J. Math. 8 (1958), 511-528.

[86] C. Menini, Jacobson's conjecture, Morita duality and related questions, J. Algebra 103 (1986), 638-655.

[87] C. Menini, Finitely graded rings, Morita duality and self-duality, Comm. Algebra 15 (1987), 1779-1797.

[72] R.W. Miller and D.R. Turnidge, Morita duality for endomorphism rings, Proc. Amer. Math. Soc. 31 (1972), 91-94.

[73] R.W. Miller and D.R. Turnidge, Some examples from infinite matrix rings, Proc. Amer. Math. Soc. 38 (1973), 65-67.

[58] K. Morita, Duality for modules and its applications to the theory of rings with minimum condition, Tokyo Kyoiku Daigaku, Ser A6 (1958), 83-142.

[69] B.J. Müller, On Morita duality, Canad. J. Math. 21 (1969), 1338-1347.

[70] B.J. Müller, Linear compactness and Morita duality, J. Algebra 16 (1970), 60-66.

[74] B.J. Müller, The structure of quasi-Frobenius rings, Canad. J. Math. 26 (1974), 1141-1151.

[84] B.J. Müller, Morita duality - A survey, Abelian Groups and Modules, Proc. Conf., Udine-Italy 1984, CISM Courses Lect. 287 (1984), 395-414.

[39] T. Nakayama, On Frobeniusean algebras. I., Ann. of Math. 40 (1939), 611-633.

[40] T. Nakayama, Note on uni-serial and generalized uni-serial rings, Proc. Imp. Acad. Tokyo 16 (1940), 285-289.

[41] T. Nakayama, On Frobeniusean algebras. II., Ann. of Math. 42 (1941), 1-21.

[72] T. Onodera, Linearly compact modules and cogenerators, J. Fac. Sci., Hokkaido Univ., Ser.I. 12 (1972), 116-125.

[66] B.L. Osofsky, A generalization of quasi-Frobenius rings, J. Algebra 4 (1966), 373-387. Erratum, 9 (1968), 120.

[91] B.L. Osofsky, Minimal cogenerators need not be unique, Comm. Algebra 19 (1991), 2071-2080.

[59] A. Rosenberg and D. Zelinsky, Finiteness of the injective hull, Math. Z. 70 (1959), 372-380.

[79] J.J. Rotman, In introduction to homological algebra, Academic Press, London-New York, 1979.

[72] F.L. Sandomierski, Modules over the endomorphism ring of a finitely generated projective module, Proc. Amer. Math. Soc. 31 (1972), 27-31.

[72b] F.L. Sandomierski, Linearly compact modules and local Morita duality, Proc. Conf. Ring Theory Utah 1971, Academic Press, London/New York, 1972, 333-346.

[85a] A.H. Schofield, Artin's problem for skew field extensions, Math. Proc. Camb. Phil. Soc. 97 (1985), 1-6.

[85b] A.H. Schofield, Representations of rings over skew fields, London Math. Soc. Lecture Note Series 92, Cambridge Univ. Press 1985.

[72] D.W. Sharpe and P. Vámos, Injective Modules, Cambridge University Press, London/New York, 1972.

[74] W. Stephenson, Modules whose lattice of submodules is distributive, Proc. London Math. Soc. (3) 28 (1974), 291-310.

[58] H. Tachikawa, Duality theorem of character modules for rings with minimum condition, Math. Z. 68 (1958), 479-487.

[73] H. Tachikawa, Quasi-Frobenius Rings and Generalizations, Lecture Notes in Mathematics 351, Springer-Verlag, Berlin-New York, 1973.

[67] Y. Utumi, Self-injective rings, J. Algebra 6 (1967), 56-64.

[77] P. Vámos, Rings with duality, Proc. London Math. Soc. 35 (1977), 275-289.

[86] J. Waschbusch, Self-duality of serial rings, Comm. Algebra 14 (1986), 581-589.

[86] E.A. Whelan, Finite subnormalizing extensions of rings, J. Algebra 101 (1986), 418-432.

[89] Weimin Xue, Artinian duo rings and self-duality, Proc. Amer. Math. Soc. 105 (1989), 309-313.

[89b] Weimin Xue, On a generalization of normalizing extensions, Comm. Algebra 17 (1989), 1093-1100.

[89c] Weimin Xue, Morita duality and artinian left duo rings, Bull. Australian Math. Soc. 39 (1989), 339-342.

[89d] Weimin Xue, Two examples of local artinian rings, Proc. Amer Math. Soc. 107 (1989), 63-65.

[89e] Weimin Xue, Morita duality and intermediate triangular extensions, Yokohama Math. J. 37 (1989), 57-59.

[90] Weimin Xue, Exact modules and serial rings, J. Algebra 134 (1990), 209-221.

[X] Weimin Xue, Artinian rings with Morita duality, preprint.

[53] D. Zelinsky, Linearly compact modules and rings, Amer. J. Math. 75 (1953), 79-90.

SUBJECT INDEX